Studies in Computational Intelligence

Volume 968

Series Editor

Janusz Kacprzyk, Polish Academy of Sciences, Warsaw, Poland

The series "Studies in Computational Intelligence" (SCI) publishes new developments and advances in the various areas of computational intelligence—quickly and with a high quality. The intent is to cover the theory, applications, and design methods of computational intelligence, as embedded in the fields of engineering, computer science, physics and life sciences, as well as the methodologies behind them. The series contains monographs, lecture notes and edited volumes in computational intelligence spanning the areas of neural networks, connectionist systems, genetic algorithms, evolutionary computation, artificial intelligence, cellular automata, self-organizing systems, soft computing, fuzzy systems, and hybrid intelligent systems. Of particular value to both the contributors and the readership are the short publication timeframe and the world-wide distribution, which enable both wide and rapid dissemination of research output.

Indexed by SCOPUS, DBLP, WTI Frankfurt eG, zbMATH, SCImago.

All books published in the series are submitted for consideration in Web of Science.

More information about this series at http://www.springer.com/series/7092

Mainak Bandyopadhyay · Minakhi Rout ·
Suresh Chandra Satapathy
Editors

Machine Learning Approaches for Urban Computing

 Springer

Editors
Mainak Bandyopadhyay
School of Computer Engineering
Kalinga Institute of Industrial Technology
(KIIT) Deemed to be University
Bhubaneswar, Odisha, India

Minakhi Rout
School of Computer Engineering
Kalinga Institute of Industrial Technology
(KIIT) Deemed to be University
Bhubaneswar, Odisha, India

Suresh Chandra Satapathy
School of Computer Engineering
Kalinga Institute of Industrial Technology
(KIIT) Deemed to be University
Bhubaneswar, Odisha, India

ISSN 1860-949X   ISSN 1860-9503   (electronic)
Studies in Computational Intelligence
ISBN 978-981-16-0934-3   ISBN 978-981-16-0935-0   (eBook)
https://doi.org/10.1007/978-981-16-0935-0

This Springer imprint is published by the registered company Springer Nature Singapore Pte Ltd.
The registered company address is: 152 Beach Road, #21-01/04 Gateway East, Singapore 189721,
Singapore

Preface

The proliferation of the Internet in the everyday life of people along with the initiatives by the governments to upgrade urban cities into smart cities has opened new challenges and areas for researchers. The urban cities are generating a lot of data on daily basis from various activities carried out in urban cities like mobility, urban transportation, traveling behavior, land-use development, consumer shopping behavior, personalized consumer behavior, health facilities, emergency services, personalized records, financial transactions, sports, etc. The data generation can be categorized into machine, organization, and people generated data. These data are captured and maintained in various storage formats by various institutions and organizations. The data ensembled from various sources help in comprehensive decision making and reveal the hidden patterns that are beneficial for improving the efficiency of various urban activities.

Advancements and popularity of machine learning techniques have encouraged researchers to gain insight into the data generated through urban activities. The activities that persist in machine learning application to urban computing can be streamlined into data collection and preparation, exploration, analysis and modeling, and generating the actionable report. The modeling of data specific to the urban region using machine learning techniques enhances the capabilities of policymakers, peoples, and organizations in predication, optimization, association analysis, cluster analysis, and classification-related applications in urban computing. The technical challenges for applying machine learning approaches to urban computing include capturing and storing heterogeneous data into various data models (semi-structured, relational, sequential, vector, and graph data models), and developing frameworks for generating machine learning models from heterogeneous data models.

The book includes contributions related to diverse application areas of machine learning and deep learning in urban computing. The book covers theoretical and foundational research, platforms, methods, applications, and tools in this area. The book presents coverage of various application areas of machine learning with respect to urban area. Some of the chapters in this book use real geographically related data.

This book will be helpful for the students, practitioners, and researchers working in the areas of GIS, urban computing, and analytics.

Bhubaneswar, India   Mainak Bandyopadhyay
                                                                   Minakhi Rout
                                                        Suresh Chandra Satapathy

Contents

Editors and Contributors

About the Editors

Dr. Mainak Bandyopadhyay did his Ph.D. in GIS and Remote Sensing from MNNIT Allahabad, Prayagraj, in 2018 after his M.Tech. from MNNIT Allahabad, Prayagraj, in 2012. He is currently working as Assistant Professor, School of Computer Engineering, KIIT Deemed to be University. He has a total of 3 years of teaching experience. His area of interest includes machine learning, urban computing and spatial algorithms. He has published a total of 10 research papers in various journals and international conferences. He is the reviewer of *International Journal of Applied Geospatial Research* (IGI Global), *Arabian Journal of Geoscience* (Springer), *International Journal of Knowledge and Systems Science* (IGI Global) and *Process Safety and Environmental Protection Journal* (Elsevier).

Dr. Minakhi Rout is currently working as Assistant Professor in the School of Computer Engineering, KIIT Deemed to be University. She has received M.Tech. and Ph.D. degree in Computer Science & Engineering from Siksha 'O' Anusandhan University, Odisha, India, in 2009 and 2015, respectively. She has more than 14 years of teaching and research experience in many reputed institutes. Her research interests include computational finance, data mining and machine learning. She has published more than 30 research papers in various reputed journals and international conferences as well as guided several M.Tech. and Ph.D. thesis. She is Editorial Member of *Turkish Journal of Forecasting*.

Dr. Suresh Chandra Satapathy did his Ph.D. in Computer Science and Engineering from JNTU, Hyderabad, after completing his M.Tech. from NIT Rourkela. He is currently working as Professor, School of Computer Engineering, KIIT Deemed to be University, Bhubaneswar, Odisha. He is also Dean of Research for Computer Engineering at KIIT Deemed to be University. He has over 30 years of teaching and research experience. He is Life Member of CSI and Senior Member of IEEE. He has been instrumental in organizing more than 59 international conferences in India and abroad as Organizing Chair and Corresponding Editor over more than 47 Book

Volumes from Springer LNCS, AISC, LNEE and SIST Series. He is quite active in research in the areas of warm intelligence, machine learning and data mining. More than 70 PG projects are supervised by him. Currently, eight scholars are pursuing Ph.D. under him. His first Ph.D. Scholar got her Ph.D. degree with GOLD medal from her University for the outstanding work done by her in the areas of warm intelligence optimization. He has over 150 research articles in various journals with SCI impact factors and SCOPUS index and also in conference proceedings of Springer, IEEE, etc. He is in Editorial Board of *IGI Global, Inderscience, Growing Science* journals, Springer *AJSE*, etc. He is the developer of a new evolutionary optimization technique called SGO and SELO. He was the recipient of Leadership in Academic award by ASSOCHEM in the year 2017.

Contributors

Sarthak Agrawal DIT University, Dehradun, Uttarakhand, India

Mainak Bandyopadhyay School of Computer Science and Engineering, Kalinga Institute of Industrial Technology Deemed-to-be-University, Bhubaneswar, Odisha, India

Anasua Banerjee School of Computer Engineering, Kalinga Institute of Industrial Technology, Bhubaneswar, Odisha, India

Gopi Battineni Medical Informatics Centre, School of Medicinal and Health Products Sciences, University of Camerino, Camerino, Italy

Vivek Bhalerao Computer Science and Engineering Department, IcfaiTech (Faculty of Science and Technology), ICFAI Foundation for Higher Education (Deemed to be University), Hyderabad, Telangana, India

H. A. Bharath RCG School of Infrastructure Design and Management, Indian Institute of Technology, Kharagpur, West Bengal, India

M. C. Chandan RCG School of Infrastructure Design and Management, Indian Institute of Technology, Kharagpur, West Bengal, India

Sayantan Ghosh School of Computer Science and Engineering, University of Kalinga Institute of Industrial Technology, Bhubaneswar, Odisha, India

Ajay Kumar Jena School of Computer Engineering, KIIT Deemed to be University, Bhubaneswar, Odisha, India

Soumen Kanrar Department of Computer Science and Engineering, Amity University Jharkhand, Ranchi, India;
Vidyasagar University, Midnapore, West Bengal, India

P. Malathi D.Y. Patil College of Engineering, Savitribai Phule Pune University, Akurdi, Pune, India

Vibham Nayak Department of Civil Engineering, Motilal Nehru National Institute of Technology Allahabad, Prayagraj, Uttar Pradesh, India

G. Nimish RCG School of Infrastructure Design and Management, Indian Institute of Technology, Kharagpur, West Bengal, India

Sandeep Kumar Panda Computer Science and Engineering Department, IcfaiTech (Faculty of Science and Technology), ICFAI Foundation for Higher Education (Deemed to be University), Hyderabad, Telangana, India

P. S. Prakash RCG School of Infrastructure Design and Management, Indian Institute of Technology, Kharagpur, West Bengal, India

Ghulam Hazrat Rezai Department of Civil Engineering, Motilal Nehru National Institute of Technology Allahabad, Prayagraj, Uttar Pradesh, India

Minakhi Rout School of Computer Engineering, Kalinga Institute of Industrial Technology, Bhubaneswar, Odisha, India

Kundan Kumar Rameshwar Saraf Senior Security Consultant, Capgemini Technology Services India Limited, Talawade, Pune, India

Suresh Chandra Satapathy School of Computer Engineering, Kalinga Institute of Industrial Technology, Bhubaneswar, Odisha, India

Anamika Sharma DIT University, Dehradun, Uttarakhand, India

Kailash Shaw D.Y. Patil College of Engineering, Savitribai Phule Pune University, Akurdi, Pune, India

Varun Singh Department of Civil Engineering, Motilal Nehru National Institute of Technology Allahabad, Prayagraj, Uttar Pradesh, India

Satyajit Swain School of Computer Engineering, Kalinga Institute of Industrial Technology, Bhubaneswar, Odisha, India

Urbanization: Pattern, Effects and Modelling

P. S. Prakash, G. Nimish, M. C. Chandan, and H. A. Bharath

Abstract Cities across the globe are expanding and increasing the paved footprint with significant disturbance in environmental factors. It is estimated that about 54% of population in the globe reside in cities and as per the World Urbanization Prospects, the urban population growth poles would be in South East Asia and Africa. The shift from primarily rural, semi-rural neighbourhood next to urban region is now extended urban societies and this is evident across multiple dimensions: as urban areas are home more than 50% of the global population, with a contribution of closer to ninety per cent of the global GDP with consumption of seventy-five per cent of energy globally. Planned sensible urbanization process provisions adequate infrastructure and basic amenities while maintaining the environmental balance. This process entails monitoring urban dynamics to understand the spatial patterns of urbanization and identify rapid growth poles. This chapter aims to understand the spatial–temporal patterns of land use dynamics in Bangalore and simulate future urbanization patterns considering a 10 km buffer using cellular automata-based model. The spatial patterns of landscape dynamics are assessed using temporal remote sensing data for three decades, acquired from Landsat repositories. Land surface temperature estimation along with understanding the changes in land use indicated a large increase in surface temperature across the region once known as green basket. This analysis also estimated the building roof area available in the city connecting the land use to estimate the solar energy potential in the region. Land use modelling indicates that the next decade, with the business as usual scenario, uncontrolled urbanization would severely erode the sustenance of natural resources (vegetation, water bodies, etc.), necessitating immediate policy measures to mitigate uncontrolled, urbanization of the city.

Keywords Urban growth · Land surface temperature · Land use · Modelling · Building extraction

P. S. Prakash · G. Nimish · M. C. Chandan · H. A. Bharath (✉)
RCG School of Infrastructure Design and Management, Indian Institute of Technology, Kharagpur, West Bengal 721302, India
e-mail: bharath@infra.iitkgp.ac.in

M. Bandyopadhyay et al. (eds.), *Machine Learning Approaches for Urban Computing*,
Studies in Computational Intelligence 968,
https://doi.org/10.1007/978-981-16-0935-0_1

1 Introduction

Cities across the globe are expanding in terms of paved footprint with significant disturbance to environmental factors. It is estimated that about 68% of the global population would reside in cities by 2050 and as per UN DESA [1]. The urban population growth poles would be in South East Asia and Africa. Shift from primarily rural or semi-rural neighbourhood next to urban region is now extended urban societies and this is evident across multiple dimensions: as urban areas are home for more than 55% of the global population. Urban region generates over 90% of the global GDP and accounts for about 75% of total global energy consumption [2, 3, 4]. Planned sensible urbanization process provisions adequate infrastructure and basic amenities while maintaining the environmental balance [5]. Also understanding the urban dynamics with the insights of the specific spatial locations growth potential would aid in evolving appropriate strategies towards the design of sustainable cities and estimating resources necessary for the future planning.

Understanding urbanization essentially requires understanding of structural growth nature of the city. Urbanization phenomenon can be well understood if it is studied at greater scale rather than small pockets of land in isolation. Several research has been carried out to understand the spatial pattern and to visualize the growth of urban area in near future [6, 7]. Prediction of land use pattern has numerous applications such as master plan generation, environmental modelling and formulating policies. Another important factor in modelling urban growth is understanding the temporal growth of the region. Remote sensing has proved to be efficient in capturing dynamically evolving urban pattern [8–10]. Increased availability of moderate resolution spatial data makes it preferred source of data for many applications. Policy makers and planning personnel need a good quality data to monitor and analyse urban land use pattern since changes in land use pattern have an impact on land surface temperature and a better understanding of surface temperature will lead to knowledge about urban climatic conditions.

The main objective of this chapter is to understand the urban growth pattern and link it to the land surface temperature as a proxy measure. Considering the research gaps, this chapter has been divided into three parts (i) understanding the dynamically evolving urban growth through land use analysis, (ii) simulate future urbanization patterns considering a 10 km buffer using cellular automata-based SLEUTH model, (iii) land surface temperature estimation along with understanding its relationship with the varying land use, and (iii) extraction of building structure footprints from high-resolution satellite imagery using soft computing techniques.

1.1 Necessity of Understanding Land Surface Temperature as a Part of Analysing Urban Pattern

Modification in landscape alters the natural energy exchange in atmosphere and affects bio-geo-chemical cycle that in turn changes the regional climate in terms of surface temperature [11, 12]. In order to understand the surface, equilibrium state and LST plays a vital role in disciplines such as urban climatology, agro-meteorology, greenhouse gas estimation, urban heat island. as a major indicator [13–17]. Alterations in climate and rise in thermal stress as a result of augmented population, industrialization, increased construction activities and optimum power generation to satisfy the residents are escalating the concentration of pollutants in the atmosphere that triggers the phenomenon of urban heat islands. Land use change dynamics in urban area is majorly dependent on buildings and manmade structures. In addition to knowledge of structural development, vertical growth of cities, accounted only by few researchers [18, 19], is also crucial in redevelopment schemes. It is paramount to understand the structural development of cities in order to formulate policy and clear objectives [20, 21].

1.2 Necessity on Extraction of Building Footprint

The growth of a city is indicated by development of infrastructures like roads and buildings. Formation of buildings shows human settlements in the cities and studying building density helps in understanding the complexities. Correlation between population growth and building density follows a linear relationship [22], proving that built-up formation is not only indicative of land use change, but also building footprints can be used to derive other parameters of urban growth. Urban land surface temperature is related to built-up volume and it is found that land surface temperature increases with the building density [23]. Building footprints of the city forms base data for modelling growth pattern in terms of land use change and it affects surface temperature.

2 Study Area and Data Used

The study has considered Bangalore city as shown in Fig. 1 as the region of interest along with a buffer of 10 km. Bangalore is located at an elevation of 900 m from mean sea level. City administrative body named Bruhat Bengaluru Mahanagara Palike (BBMP) consists of 198 wards as administrative divisions. Geographically, Bangalore city lies between 12° 47′ N and 13° 09′ N and 77° 27′ E and 77° 47′ E. During late 1995 and early 2000, Bangalore experienced exponential growth as an outcome

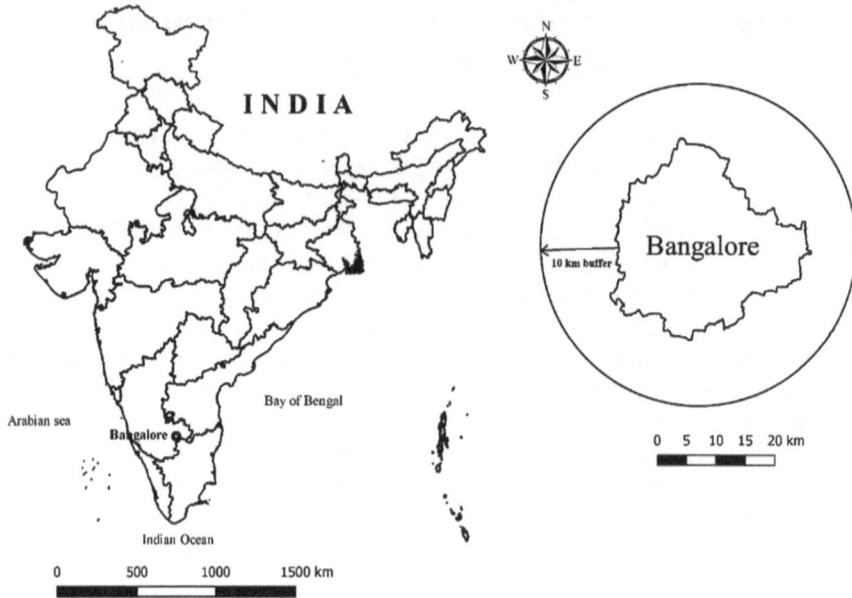

Fig. 1 Study area—Bangalore city and buffer zone of 10 km

of IT sector establishment and industrialization that lead to rapid land use change in the outskirts of the city. Table 1 gives the details about various spatial data sets used to carry out analysis and modelling.

Table 1 Specifications of spatial data used in the study

Purpose	Satellite sensor	Resolution Spatial and spectral	Year range considered
Land use analysis	Landsat ASTER	Bands: NIR, Red, Blue—30 m	1992, 1999, 2009, 2017
Land surface temperature	Landsat	Bands: 6 (TM, ETM +), 10 (OLI); Thermal bands: 30 m	1992, 1999, 2009, 2017
Building footprint extraction	Triple Sat	Pan: 0.8 m Multispectral: 3.2 m Bands: Blue, Green, Red	2018
Slope, hillshade	Aster Dem	Spatial: 30 m	2012/2018
Exclusion data	Google earth		2002–2017

3 Method

Acquired satellite data is preprocessed and geo-registered. The study area boundary is digitized from topo sheets. Satellite data was then cropped pertaining to study area for further processing. Slope map and hillshade maps are generated directly from elevation data and vector layers (transportation and excluded) required for model are manually digitized. A false colour composite prepared from satellite imagery is subjected to supervised classification for preparing land use map consisting of classes, namely urban, water, vegetation and others. Accuracy is assessed using error matrix and kappa statistics. An urban growth model based on cellular automata called SLEUTH is used to model the future possible changes in urban growth. SLEUTH uses layers of information over several years of data and predicts future urban growth. A high-resolution satellite imagery is utilized to extract building footprints from the study area, which acts as supporting data for urban growth model validation. Various machine learning techniques were under used and SVM was found to be closer to better building footprint extraction. Therefore, support vector machine, a machine learning technique, is used to extract building from satellite imagery. Understanding the relationship between urban growth and climatic parameter such as surface temperature is performed. Thermal band of extracted satellite imagery is helpful in deriving land surface temperature from empirical methods.

3.1 Land Use Analysis

Land use analysis was performed using the standard combination of false colour composite for all the heterogenous patches. Land use analysis was performed as described in Ramachandra et al. Accuracy in terms of users, producers, overall accuracy and kappa coefficients was estimated to understand how closely these maps were in agreement with real-time scenario.

3.2 Spatial Metrics and Urban Expansion Analysis

The study area wad divided into gradients and zones for regional level analysis. It was divided into concentric circles of 1 km and four cardinal directions. This provides an essential input of understanding every pocket of the urban landscape growth. Based on literature reviewed [24–26], three metrics were used on the study: number of urban patches (NP), normalized land shape index of urban patch (NLSI) and clumpiness index (CLUMPY) were computed using Fragstats [27].

3.3 Urban Growth Indices

Two indices annual increase index (AII) and annual growth rate index (AGRI) were adapted to quantify urban growth rate during study period [28]. AII is used to measure annual changes of urban areas, while AGRI eliminates size aspect of a city and suitable for comparison of city at different period. Two indices are given in Fig. 2; Table 2.

3.4 Land Use Modelling Using SLEUTH

Data is preprocessed as per the requirement of SLEUTH model including data preparation and standardization. SLEUTH analysis has three phases that includes the data phase of input, phase of verification, and calibration of the model consists of three phases: input data phase, verification and calibration phase and the final phase is the prediction. Phase of verification is to confirm the data preparation in correct method. Calibration and prediction is carried out to improve the learning capability of the model to predict the future scenario to achieve best-fit statistic of each coefficient used in analysis. Calibration is carried out in three different modes: coarse, fine and final with iterations numbers considered based on previous literature [29]. Top Lee–Salee metric is selected to obtain five unique coefficient values responsible for urban growth.

3.5 Building Footprint Extraction

Machine learning techniques are extensively used among various methods of information extraction from satellite imageries and satisfies essential accuracy requirement [30, 31]. In this study, a pan-sharpened multispectral imagery having 80 cm spatial resolution and consists of blue, green and red bands is used to extract building footprints. Training data sets are captured from the imagery as building and non-building classes, building class encompasses multiple types of structures such as individual house, factories, apartments, office or commercial buildings, and non-building class consists of the waterbody, open spaces, roads, vegetation, etc. Enough care is taken in the process of signature capture to ensure minimal occurs in terms of misjudgment or overlapping classes. Visual reference from Google Earth and ground truth verification have been performed at chosen places to make sure that training stage is efficient enough to predict classes correctly thought the study area. A machine learning method namely support vector machine (SVM) is employed to classify. SVM is trained using training data and parameters required for radial basis kernel function that is given to predict classes of unknown pixels in the imagery. The predicted building footprints are subjected to a morphological closing operation to make it closer to reality.

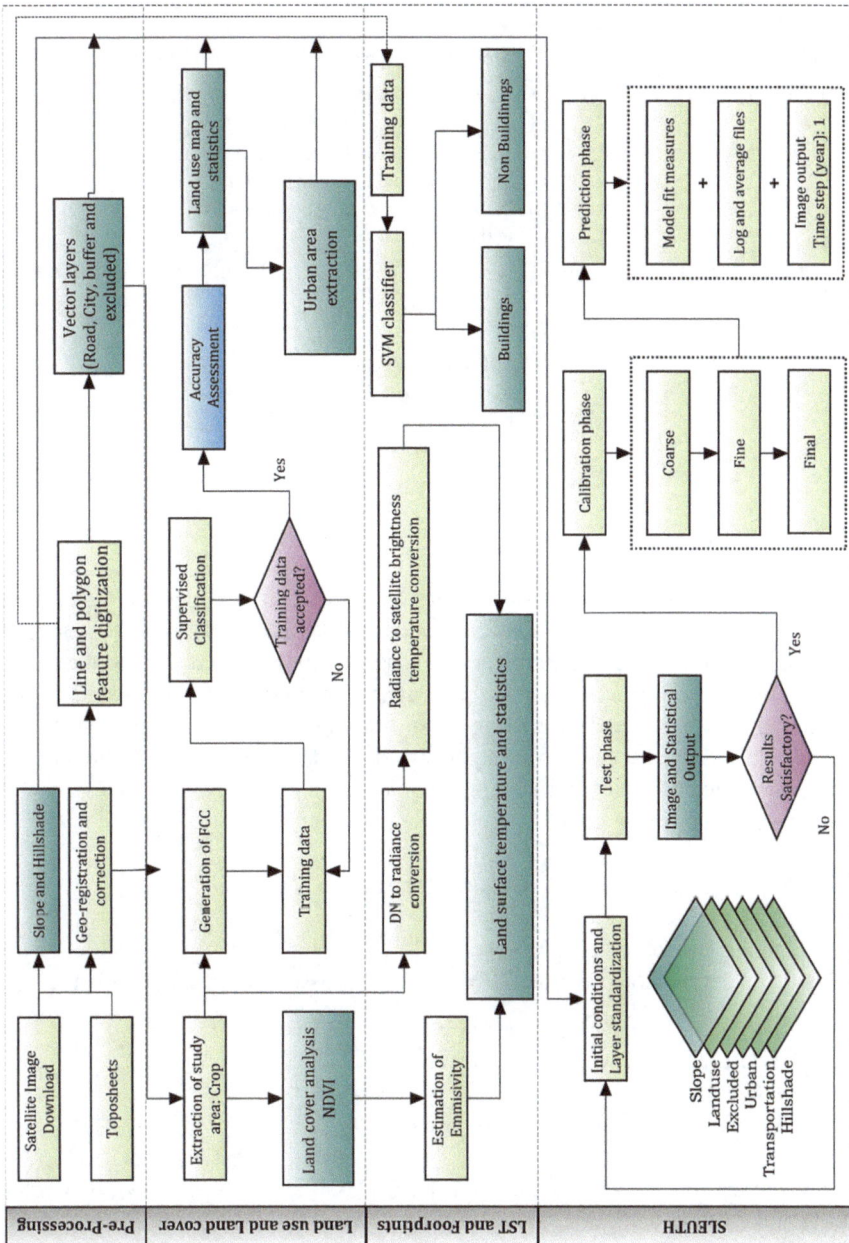

Fig. 2 Flow chart showing integrated model approach

Table 2 Urban growth indices used

	Indicators	Metric/index type and formula	Range
1	Annual increase index—AII	$AII = \frac{U_{\text{final}} - U_{\text{start}}}{t}$ U_{start} and U_{final} refer to urban area at starting and end of time period considered for analysis, t being the time span between start and end in years	AII > 0, without limit
2	Annual growth rate index—AGRI	$AGRI =$ $\left[\left(\frac{U_{\text{final}}}{U_{\text{start}}} \right)^{1/t} - 1 \right] * 100 \%$ AII (km^2 per year) and AGRI (%) are annual urban growth extent indicator and annual urban growth rate respectively	$0 \leq AGRI \leq 100\%$

3.5.1 Support Vector Machine

It is a special type of supervised machine learning method; it classifies data by creating hyperplanes in a multidimensional feature space. Classifier robustness comes from maximizing space between hyperplane and corresponding support vectors of classes. A small number of training data sets are capable of producing good results as the classification is based on considering those training samples that are close to the margin. Hence, the success of an algorithm depends on how good training data, which is unique and well separated to other classes.

Let us take a training sample

$$\{(S_i d_i)\} \qquad \text{for } i = 0 \text{ to } N$$

N: Training sample size, $d_i = +1$ or -1.

Based on class belongingness, SVM classifies classes by using hyperplanes.

$S_i^T Y_i + b \geq 0$	for $d_i = +1$
$S_i^T Y_i + b \leq 0$	for $d_i = -1$

Y_i Input vector, S AdjusTable weight, b Parameter bias

Above-mentioned SVM method is a linear classification type shown by Cortes and Vapnik [32]. In addition to these nonlinear types of SVM classification as given in Hsu et al. [33] SVM is capable of mapping data into multidimensional space. References [34, 35] and Huang et al. demonstrated applicability of various kernel functions of SVM classification.

Linear: $G(Y, Y_i)$	$= Y^T Y_i$
Polynomial: $G(Y, Y_i)$	$= (\gamma Y^T Y_i + b)^\rho, \ \gamma > 0$
Radial basis function: $G(Y, Y_i)$	$= \exp(-\gamma \|Y - Y_i\|^2), \ \gamma > 0$
Sigmoid: $G(Y, Y_i)$	$= \tanh(\gamma Y^T Y_i + r)$

Y input vector, Y_i feature space vector, ρ degree of polynomial, b bias term for polynomial and sigmoid; γ gamma term of kernels (polynomial, radial basis function and sigmoid)

3.5.2 Accuracy Assessment of the Building Footprint Extraction

Captured signatures of two classes are divided into a training data set and testing data set in 70:30 ratio. Here, 70% of signatures are used for training the classifier and 30% is used for testing classifier accuracy. The confusion matrix is prepared contains accuracy, sensitivity and specificity. Sensitivity is the measure of correctly classified buildings and specificity denotes wrongly classified buildings.

$$\text{Sensitivity} = X/(X + Z); \ \text{Specificity} = E/(Y + E);$$
$$\text{Accuracy} = (X + E)/(X + Y + Z + E)$$

where $X =$ True Negative; $Z =$ False Negative; $E =$ True Positive; $Y =$ False positive.

3.6 Land Surface Temperature Analysis

Thermal remote sensing is carried out in the range of 10.4–12.5 μm. The sensor captures the data corresponding to reflectance/emittance in the form of digital numbers, and to extract quantitative information from these, it has to be converted into spatial radiance and top-of-atmosphere temperature. The study incorporates single window algorithm, one of the most common yet efficient methods that uses only one thermal band to estimate land surface temperature. It is a three-step process as shown.

3.6.1 Calculation of At-Satellite Brightness Temperature

DN values were converted into spectral radiance by using gain and offset (as provided in metadata) as denoted in Eq. 1

$$L_\lambda = (\text{Gain} * \text{DN}) + \text{Offest} \tag{1}$$

Here, $L_\lambda =$ spectral radiance.

Post this at-satellite brightness temperature was obtained using Eq. 2

$$T_B = \frac{K_2}{\ln\left(\frac{K_1}{L_\lambda} + 1\right)} \tag{2}$$

Here, T_B = brightness temperature, K_1, K_2 are constants.

3.6.2 Estimation of Emissivity

Emissivity is a vital parameter for estimation of LST. It can be derived by multiple methods and for this study NDVI threshold method was used that distinguishes emissivity because of normalized difference vegetation index (NDVI) values. CMAP, 2018, was utilized to get the emissivity values for classes: water, soil and vegetation; while for mixed pixel (soil and vegetation), proportion of vegetation (P_v) was estimated (Eq. 3) followed by emissivity as shown in Eq. 4.

$$P_V = \left(\frac{\text{NDVI} - \text{NDVI}_S}{\text{NDVI}_V - \text{NDVI}_S}\right)^2 \tag{3}$$

NDVI can be defined as the amount of vegetation present in the region. Here, NDVI_S and NDVI_V denote NDVI thresholds for soil and vegetation, respectively.

$$\varepsilon_{SV} = \varepsilon_V P_V + \varepsilon_S (1 - P_V) + C \tag{4}$$

Here, ε_{SV} = emissivity of soil and vegetation; ε_V = emissivity of vegetation; ε_S = emissivity of soil; C is a constant that defines surface characteristics and can be estimated using Eq. 5

$$C = (1 - \varepsilon_S)\varepsilon_V F(1 - P_V) \tag{5}$$

F = geometrical factor (0.55).

3.6.3 Quantification of LST

Every feature on the earth's surface that has temperature above zero kelvin tends to emit the radiations it has absorbed. Thus, LST is not just the brightness temperature but it is an integration of at-satellite temperature and emissivity. It can be estimated using Eq. 6

$$\text{LST} = \frac{T_B}{1 + \left(\frac{\lambda T_B}{\rho} X \ln(\varepsilon)\right)} \tag{6}$$

Here, λ denotes the wavelength at which maximum relative response is observed; $\rho = \frac{hc}{\sigma} = \left(1.438 \times 10^{-2}\right)$ mK (h is Plank's constant, c is speed the of light and σ is Stefan Boltzmann constant).

4 Results

4.1 Land Use Analysis

Results and statistics of land use analysis are shown in Table 3 and Fig. 3. Bangalore has accounted for 348% increase in urban cover over the past two and half decades. Especially in the periphery of the administrative area, infrastructure and residential development have occurred at the cost of agricultural land, lakes, lakebed, wetlands, etc. This phenomenon of rapid land use change within a short span of time can be visualized from Fig. 3. Statistics show steep decline of vegetation from 17.01% (1992) to 5.79% (2017) and waterbody has from 2.42% (1992) to mere 0.7% (2017). Bangalore in the early 90 has had around 180 waterbodies and merely a few in comparison with present scenario. This decline in waterbodies is either due to converting to other forms of land use or lakes have dried over the period due to poor maintenance and encroachment/debris filling. Overall accuracy varied from 90 to 94%, whereas kappa coefficient obtained varied from 0.78 to 0.87.

Table 3 Land use classification and accuracy assessment statistics for Bangalore region

Year	Urban (%)	Vegetation (%)	Water (%)	Others (%)	OA (%)	K
1992	5.47	17.01	2.42	75.10	94	0.87
1999	8.00	9.26	1.94	80.80	94	0.83
2009	13.48	11.48	0.78	74.26	92	0.82
2017	24.53	5.79	0.70	68.98	90	0.78

OA Overall accuracy; *K* Kappa coefficient

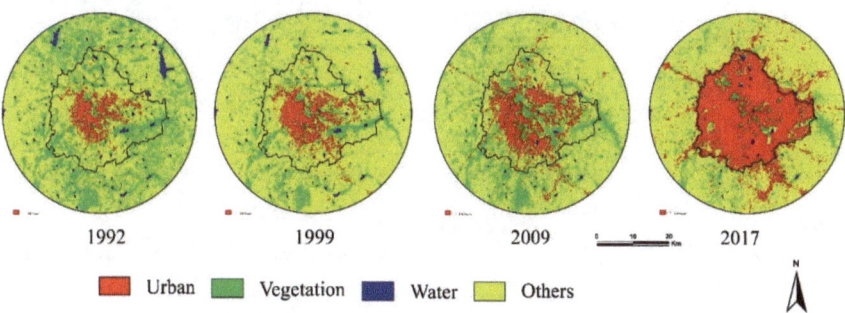

Fig. 3 Land use dynamics of Bangalore, 1992–2017

Spatial metrics and urban expansion analysis

Spatial metrics were calculated using Fragstats. The metric NP was calculated both circle wise and direction wise. NP indicated that the higher number of any specific class patch would be indicative of sprawl uncluttered mixed land uses. Values closed to one indicate a single dominant patch. The analysis of the results for this metric indicates that core has reached the threshold of urban growth, whereas the periphery and the outer circles have a dispersed growth. This is common in all the four directions. The outer circles of the city region indicate the growth was paced post 2010, whereas inner circles in the city regions had a periodical growth of patches in 1990s and started aggregating to form urban patch post 1999. The results are represented in Fig. 4. NLSI is a measure of conversion of land to a class or several mixed classes based on shape. Value of NLSI is 0–1. 1 indicates the regions have complex shape indicating the mixed class, whereas values closer to 0 indicate that the region has intense a particular growth of class indicating a compact land use change. Analysis revealed that all four zones show lesser value of NLSI in 2017 when compared to 1992 in the outer circles as depicted in Fig. 5, whereas centre has values close to 0.

Clumpiness in another metric that validates the aggregation and disaggregation but for adjacent urban patches, therefore, is essential metric in understanding the regional growth. Referring to Fig. 6, in 2017, the values ranging from 0.8 to 1 (circles 1–13) indicating growth are very complex and patches are maximally aggregated to form wide urban monotype landscape. Graph also depicts values tending towards 0 in all directions (circles 15–26), indicating less compact growth or maximum disaggregation.

Table 4 depicts temporal rate of change in annual increase index and annual urban growth rate index. The annual changes of urban areas and growth rate for the

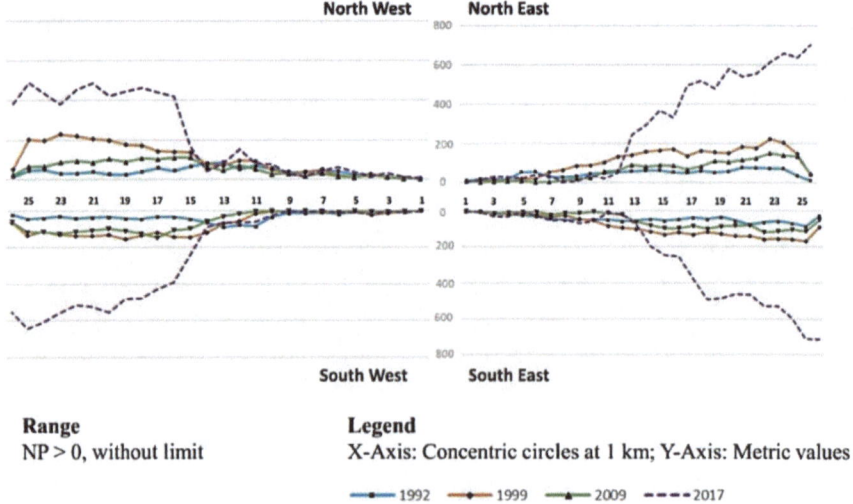

Fig. 4 Number of patch metric for Bangalore during 1992–2017

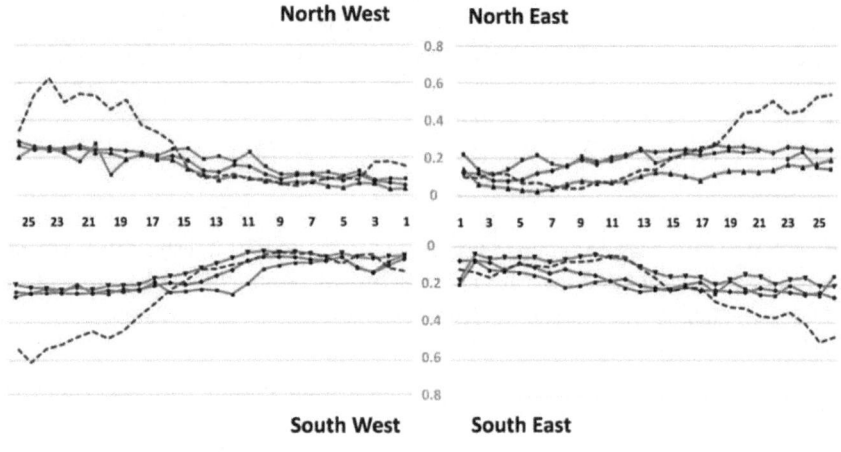

Range **Legend**
$0 \leqq NLSI \leqq 1$ X-Axis: Concentric circles at 1 km; Y-Axis: Metric values

1992 1999 2009 2017

Fig. 5 Normalized landscape shape index for Bangalore during 1992–2017

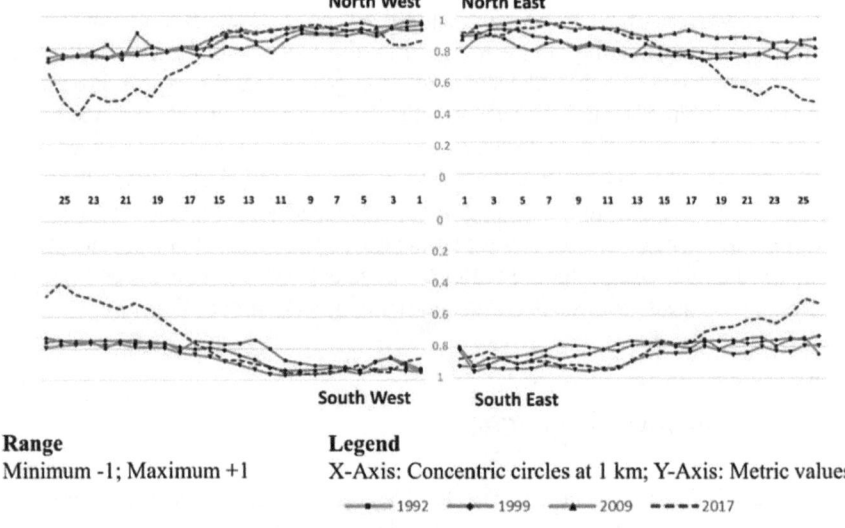

Range **Legend**
Minimum -1; Maximum +1 X-Axis: Concentric circles at 1 km; Y-Axis: Metric values

1992 1999 2009 2017

Fig. 6 Clumpiness metric for Bangalore during 1992–2017

Table 4 Urban growth indices for Bangalore region from the year 1992 to 2025

Period	1992–1999	1999–2009	2009–2017	2017–2025 (Predicted)
AII (km^2)	0.36	0.54	1.38	4.14
AGRI (%)	5.58	5.35	7.77	11.28

period 2009–2017 were recorded as highest with 1.38 and 7.77, respectively, suggest Bangalore witnessed rapid increase in urban growth along with highest rate due to industry-oriented policy measures. Values also correspond to results obtained from land use analysis and visualized in Fig. 3.

4.2 Results of Modelling Urban Land Use Change Using SLEUTH

SLEUTH model based on CA was used to visualize the urban growth for the year 2025 as shown in Fig. 8. Using metrics such as Lee–Salee, the final coefficient values described very less diffusion and has less affinity to breed and slope values. There was resistance to growth influenced by slope in various parts especially the regions closer to state highways, national highways and highly slopy areas closer to Indian Institute of science, and surrounding areas. Medium road gravity contribution for land use change was observed for the values of spread suggested that the regions around the city roads are high potential for such drastic change in land use. Lee-Salee metric has a value of 0.8 indicating the model has been well trained [36–39] and success of model iterations. It is evident that Bangalore has been intensely following the urban pattern as per its historical trends and the other category in the land use would be the most transitioning category to urban land use. Urban cover would increase to 57.62% of the total area including buffer from the current urban land use of 24.33% insisting the necessity of urgent vision and policy interventions (Fig. 7).

4.3 Land Surface Temperature

Land surface temperature was extracted as described and the results are as shown in Fig. 9. The results indicate that in last three decades, the mean surface temperature had seen an upsurge of 8 °C. Land use maps were used to extract each class and LST was estimated. Statistical analysis concerning each class was carried out as one of the parameter is represented in Table 5 and it was found that others and urban category correlate to higher temperatures followed by vegetation and water bodies. Other categories that include unpaved areas, open agricultural fields, mining sites, etc. have experienced a rise in temperature from 34.57 to 42.13 °C. Development in

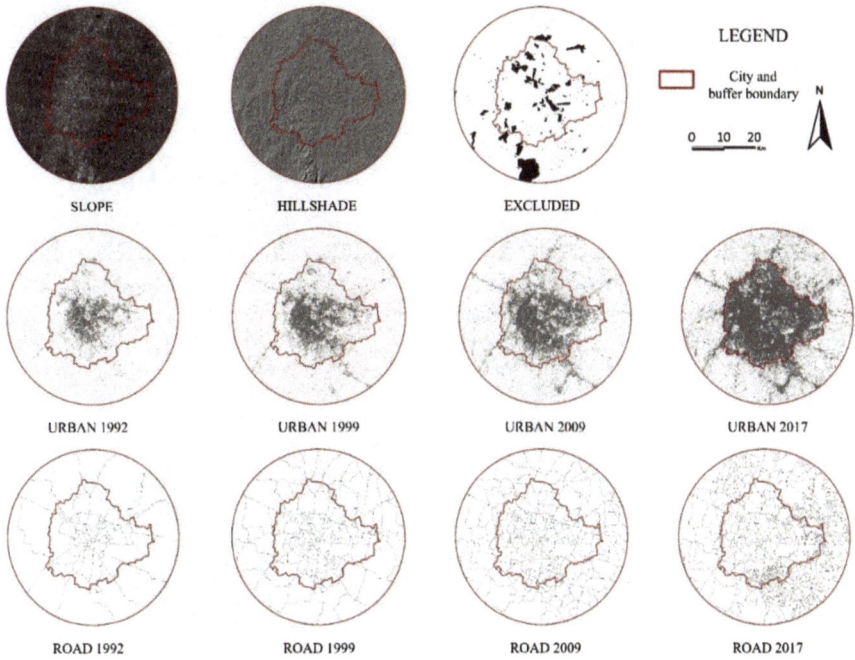

Fig. 7 Input layers used for SLEUTH model

Bangalore in terms of paved surfaces took place at the cost of vegetation and water bodies, reduction in both the categories, as well as reduction in depth and quality for water bodies has led to rise in mean surface temperature of both the classes. Airfields in Jakkur, HAL and Kempegowda International airport show the highest temperature (44–51 °C) while green areas such as Bangalore golf course, dense vegetation patch in Indian Institute of Science, Cubbon park, shows less temperatures (33–35 °C). Water bodies in the form of lakes and tanks show lowest temperature (30–33 °C). Intra-class variability was also estimated and understood by quantifying coefficient of variation (COV) and it was found that water body had the highest value. This can be inferred due to the presence of contamination in water bodies along with variation in depths. Others, urban and vegetation categories show moderate value signifying not much difference in species of trees, open areas and type of construction material used.

4.4   Building Footprint Extraction

Machine learning classifier SVM gives predicted classes as continues sets of pixel values are divided into two classes using on a threshold, which is determined by careful observation of the predicted data. Morphological fill operation helps to fill

Fig. 8 Predicted urban growth for Bangalore for the year 2025

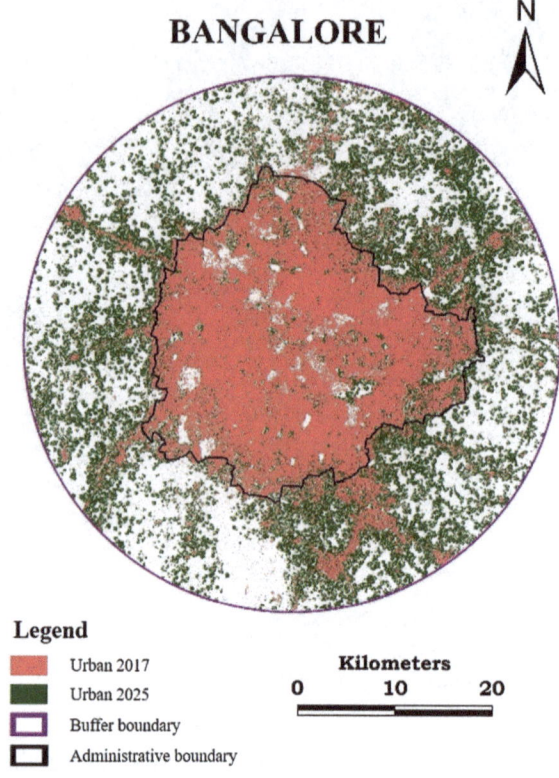

the unwanted gaps in the output and hence to obtain true building classes. Figure 10 shows the subset of satellite imagery and corresponding classified map. Details of the area obtained from the study area are given in Table 6.

5 Conclusion

The communication effectively signifies the use of remote sensing and GIS applications in urban growth and climate. It correlates the changing urban areas with surface temperature category/class wise. The city of Bangalore considered a garden city that maintains a healthy climate but it is now trending towards a non-sustainable city with poor living condition in terms of environment. Results show that there has been a rise in urban area by 348.44% in past two and half decades that contributed for an overall rise of 7.96 °C in the mean surface temperature of the city. Spatial metrics helped in visualizing sprawling pattern and its effect on natural resources. Metrics also revealed densification in core city area contributing to increased land surface temperature. The increased thermal discomfort of the city can be inferred to reduction

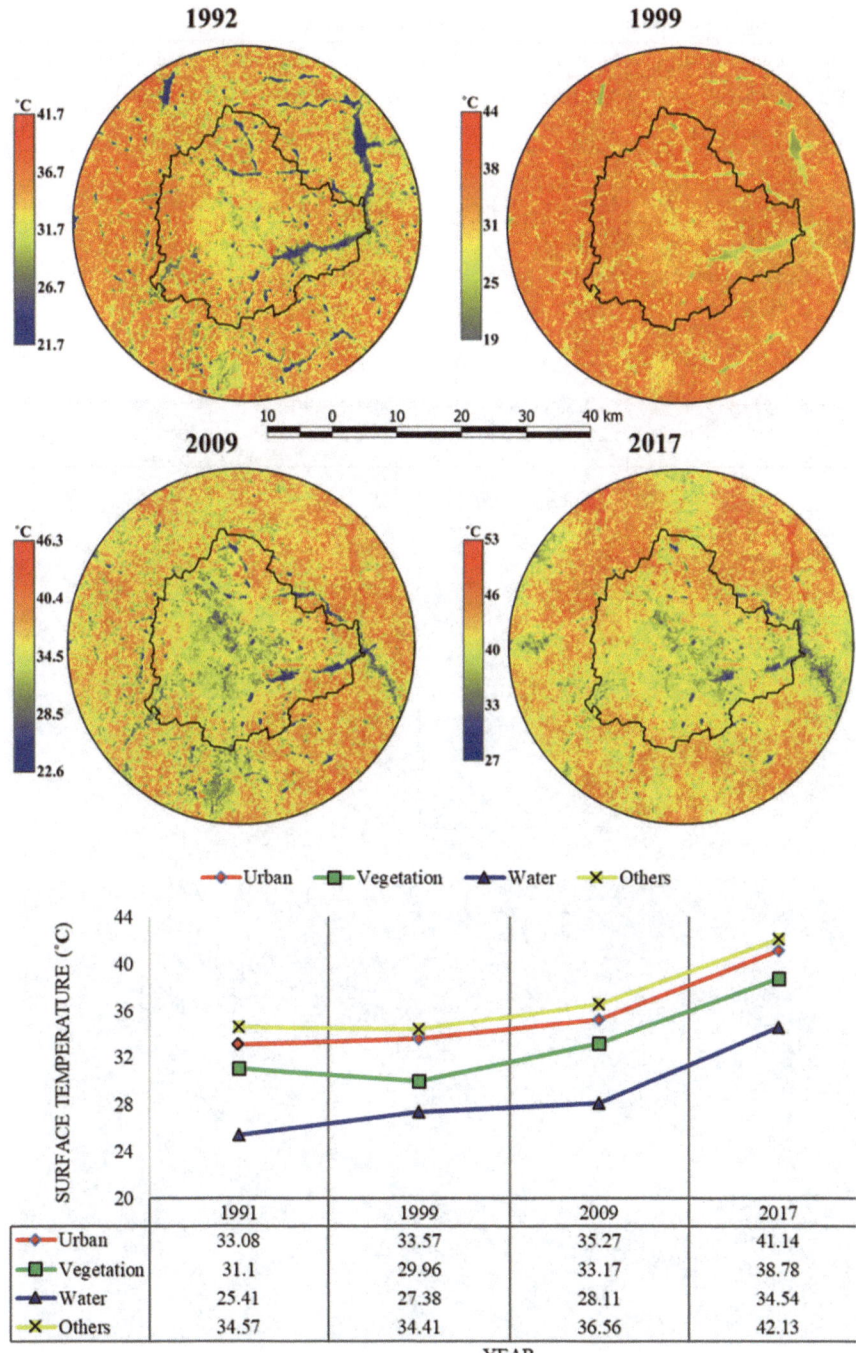

Fig. 9 Land surface temperature for greater Bangalore with 10 km buffer

Table 5 Temperature statistics for Bangalore with 10 km buffer

Year	Land use class	Mean Temp. (°C)	Year	Land use class	Mean Temp. (°C)
1992	Urban	33.07	**2009**	Urban	35.26
	Vegetation	31.09		Vegetation	33.16
	Water	25.40		Water	28.11
	Others	34.57		Others	36.56
1999	Urban	33.56	**2017**	Urban	41.14
	Vegetation	29.95		Vegetation	38.77
	Water	27.37		Water	34.53
	Others	34.41		Others	42.12

Fig. 10 Subset of satellite imagery and extracted buildings superimposed on imagery

Table 6 Area details of building footprint extracted	Classes	Buildings	Non-buildings	Total
	Area	210.15	501.74	711.60

in vegetation cover and water bodies for development of impervious surfaces. The communication shows the linkage of water bodies and vegetation cover in regulating the surface temperature, thus, soothing the microclimate of the region. Model-based approach adopted using earth observation data predicted changes in land use in the outskirts of the Bangalore city considering past growth rate by spatial means.

SLEUTH model was successfully tested, calibrated in three different phases with validation using fit statistics. Regressed values of urban pixel population, urban edges, urban clusters and urban cluster size showed values greater than 0.8 and Lee–Salee greater than 0.4 which were treated as satisfactory and justifying the validation of real-time scenario. Further, modelled results accounted the increase in urban area of 595.42 km^2 within a span of eight years (2017–2025). The successful demonstration of SLEUTH model justifies adaption of cellular automata (CA)-based self-organizing transition rule helps capturing realistic near future prediction of urban growth. However, researchers suggest integrating better optimization techniques with SLEUTH to explore improved performance of models with high level accuracy [40–42]. Proposed machine learning method extracts buildings from the high-resolution satellite imagery efficiently to quantity urban growth and helps in validation of prediction models. The building details with urban growth pattern and surface urban temperature may be used to understand the changes and improving the policy towards developing sustainable cities.

Though the analysis could draw essentially the understanding between urban growth and proxy parameters such as land surface temperature to link it to climate change, it would be more successful if many more proxy parameters are used to define the change. The data used here is 30 m for understanding the urban growth pattern. Since the urban growth is now centred around the infrastructure necessities, using a better spatial resolution data would be essential. SVM is one of the well-known machine learning techniques. But improvised techniques such as using Nuerals would provide extremely important information for the planners of city.

Acknowledgements We acknowledge the funding from Science and Engineering Research Board, Government of India and Department of Science and Technology, Government of West Bengal. We also acknowledge the infrastructural support from Indian Institute of Technology Kharagpur and data sources of open platforms of Landsat.

References

1. UN DESA, 2018 Revision of world urbanization prospects (2018). Accessed on 27 Sept 2018. Retrieved from https://www.un.org/development/desa/publications/2018-revision-of-world-urbanization-prospects.html

2. M. Azhar Khan, M. Zahir Khan, K. Zaman, L. Naz, Global estimates of energy consumption and greenhouse gas emissions. Renew. Sustain. Energy Rev. **29**, 336–344 (2014)
3. F. Krausmann, S. Gingrich, N. Eisenmenger, K.-H. Erb, H. Haberl, M. Fischer-Kowalski, Growth in global materials use, GDP and population during the 20th century. Ecol. Econ. **68**(10), 2696–2705 (2009)
4. T.V. Ramachandra, G. Kulkarni, H.A. Bharath, S.S. Han, GHG emissions with the mismanagement of municipal solid waste: case study of Bangalore, India. Int. J. Environ. Waste Manage. **20**(4), 346 (2017)
5. M.M.P. Rana, Urbanization and sustainability: challenges and strategies for sustainable urban development in Bangladesh. Environ. Dev. Sustain. **13**(1), 237–256 (2010)
6. H.A. Bharath, M.C. Chandan, S. Vinay, T.V. Ramachandra, Modelling urban dynamics in rapidly urbanising Indian cities. Egypt. J. Remote Sens. Space Sci. **21**(3), 201–210 (2017)
7. E. Vaz, J.J. Arsanjani, Predicting urban growth of the greater toronto area—coupling a markov cellular automata with document meta-analysis. J. Environ. Inf. **25**(2), 71–80 (2015)
8. H.A. Bharath, S. Vinay, M.C. Chandan, B.A. Gouri, T.V. Ramachandra, Green to gray: Silicon Valley of India. J. Environ. Manage. **206**, 1287–1295 (2018)
9. S.I. Musa, M. Hashim, M.N.M. Reba, A review of geospatial-based urban growth models and modelling initiatives. Geocarto Int. **32**(8), 813–833 (2017)
10. R. Padmanaban, A.K. Bhowmik, P. Cabral, A. Zamyatin, O. Almegdadi, S. Wang, Modelling urban sprawl using remotely sensed data: a case study of Chennai city, Tamilnadu. Entropy **19**(4), 163 (2017)
11. S. Bharath, K.S. Rajan, T.V. Ramachandra, Land surface temperature responses to land use land cover dynamics. Geoinfrmatics Geostatistics: An Overview **1**(4), 1–10 (2013)
12. M. Jin, J. Li, C. Wang, R. Shang, A practical split-window algorithm for retrieving land surface temperature from Landsat-8 data and a case study of an urban area in China. Remote Sens. **7**(4), 4371–4390 (2015)
13. M.C. Anderson, J.M. Norman, W.P. Kustas, R. Houborg, P.J. Starks, N. Agam, A thermal-based remote sensing technique for routine mapping of land-surface carbon, water and energy fluxes from field to regional scales. Remote Sens. Environ. **112**, 4227–4241 (2008)
14. S. Khandelwal, R. Goyal, N. Kaul, A. Mathew, Assessment of land surface temperature variation due to change in elevation of area surrounding Jaipur, India. Egypt. J. Remote Sens. Space Sci. **21**(1), 87–94 (2018)
15. Z.L. Li, F. Becker, Feasibility of land surface temperature and emissivity determination from AVHRR data. Remote Sens. Environ. **43**(1), 67–85 (1993)
16. Z.-L. Li, B.-H. Tang, H. Wu, H. Ren, G. Yan, Z. Wan, J.A. Sobrino, Satellite-derived land surface temperature: current status and perspectives. Remote Sens. Environ. **131**, 14–37 (2013)
17. T.J. Schmugge, F. Becker, Remote sensing observations for the monitoring of land-surface fluxes and water budgets, in *Land Surface Evaporation*. ed. by T.J. Schmugge, J.C. André (Springer, New York, USA, 1991), pp. 337–347
18. J. Lin, B. Huang, M. Chen, Z. Huang, Modeling urban vertical growth using cellular automata—Guangzhou as a case study. Appl. Geogr. **53**, 172–186 (2014)
19. L. Salvati, M. Zitti, A. Sateriano, Changes in city vertical profile as an indicator of sprawl: evidence from a Mediterranean urban region. Habitat Int. **38**, 119–125 (2013)
20. M. Batty, K.W. Axhausen, F. Giannotti, A. Pozdnoukhov, A. Bazzani, M. Wachowicz et al., Smart cities of the future. Eur. Phys. J. Spec. Top. **214**(1), 481–518 (2012)
21. S. Campbell, Green cities, growing cities, just cities? Urban planning and the contradictions of sustainable development. J. Am. Plann. Assoc. **62**(3), 296–312 (2015)
22. L. Tomas, L. Fonseca, C.M. Almeida, F. Leonardi, M. Pereira, Urban population estimation based on residential buildings volume using IKONOS-2 images and lidar data. Int. J. Remote Sens. **37**(1), 1–28 (2016)
23. M. Ranagalage, R.C. Estoque, H.H. Handayani, X. Zhang, T. Morimoto, T. Tadono, Y. Murayama, Relation between urban volume and land surface temperature: a comparative study of planned and traditional cities in Japan. Sustainability (Switzerland) **10**(7), 1–17 (2018)

24. F. Kong, H. Yin, N. Nakagoshi, P. James, Simulating urban growth processes incorporating a potential model with spatial metrics. Ecol. Ind. **20**, 82–91 (2012)
25. J. Sadhana, K. Divyani, R. Ram Mohan, B. Wietske, Spatial metrics to analyze the impact of regional factors on pattern of urbanisation in Gurgaon, India. J. Indian Soc. Remote Sens. **39**(2), 203–212 (2011)
26. H. Sudhira, T.V. Ramachandra, K.S. Jagadish, Urban sprawl: metrics, dynamics and modelling using GIS. Int. J. Appl. Earth Obs. Geoinf. **5**(1), 29–39 (2004)
27. K. McGarigal, B. Marks, *Fragstats—spatial pattern analysis program for quantifying landscape structure*. Forest Science Department, Oregon State University. Retrieved from https://www.fs.fed.us/pnw/pubs/pnw_gtr351.pdf
28. W. Wu, S. Zhao, C. Zhu, J. Jiang, A comparative study of urban expansion in Beijing, Tianjin and Shijiazhuang over the past three decades. Landscape Urban Plann. **134**, 93–106 (2015)
29. E.A. Silva, K.C. Clarke, Calibration of the SLEUTH urban growth model for Lisbon and Porto, Portugal. Comput. Environ. Urban Syst. **26**(6), 525–552 (2002)
30. O. Benarchid, N. Raissouni, S. El Adib, A. Abbous, A. Azyat, N.B. Achhab, A. Chahboun, Building extraction using object-based classification and shadow information in very high-resolution multispectral images, a case study: Tetuan, Morocco. Can. J. Image Process. Comput. Vision **4**(1), 1–8 (2013)
31. L. Bruzzone, L. Carlin, A multilevel context-based system for classification of very high spatial resolution images. IEEE Trans. Geosci. Remote Sens. **44**(9), 2587–2600 (2006)
32. C. Cortes, V. Vapnik, Support-vector networks. Mach. Learn **20**(3), 273–297 (1995)
33. C.W. Hsu, C.C. Chang, C.J. Lin, A practical guide to support vector classification (2003). Retrieved from https://www.csie.ntu.edu.tw/~cjlin/papers/guide/guide.pdf
34. T. Kavzoglu, I. Colkesen, A kernel functions analysis for support vector machines for land cover classification. Int. J. Appl. Earth Obs. Geoinf. **11**(5), 352–359 (2009)
35. X. Yang, Parameterizing support vector machines for land cover classification. Photogram. Eng. Remote Sens. **77**(1), 27–37 (2011)
36. H.F. Kuo, K.W. Tsou, Modeling and simulation of the future impacts of urban land use change on the natural environment by SLEUTH and cluster analysis. Sustainability **10**(1), 72 (2017)
37. G. Nimish, M.C. Chandan, H.A. Bharath, Understanding current and future land use dynamics with land surface temperature alterations: a case study of Chandigarh. ISPRS Ann. Photogrammetry, Remote Sens. Spat. Inf. Sci. IV-5, 79–86 (2018)
38. T. Osman, P. Divigalpitiya, T. Arima, Using the SLEUTH urban growth model to simulate the impacts of future policy scenarios on land use in the Giza Governorate, Greater Cairo Metropolitan region. Int. J. Urban Sci. **20**(3), 407–426 (2016)
39. Y. Sakieh, B.J. Amiri, A. Danekar, J. Feghhi, S. Dezhkam, Simulating urban expansion and scenario prediction using a cellular automata urban growth model, SLEUTH, through a case study of Karaj City Iran. J. Hous. Built Environ. **30**(4), 591–611 (2015)
40. Y. Feng, Y. Liu, X. Tong, M. Liu, S. Deng, Modeling dynamic urban growth using cellular automata and particle swarm optimization rules. Landscape Urban Plann **102**(3), 188–196 (2011)
41. J. Jafarnezhad, A. Salmanmahiny, Y. Sakieh, Subjectivity versus objectivity: comparative study between brute force method and genetic algorithm for calibrating the SLEUTH urban growth model. J. Urban Plann. Dev. **142**(3), 05015015 (2015)
42. K.C. Clarke, LJ Gaydos, Loose-coupling of a cellular automaton model and GIS: Long-term growth prediction for the San Francisco and Washington/Baltimore. Int. J. Geogr. Inf. Sci. 12, 699–714 (1998)

A Spatiotemporal Accessibility Analysis of Bus Transportation Facility in the National Capital Territory (NCT) of Delhi, India

Ghulam Hazrat Rezai, Varun Singh, and Vibham Nayak

Abstract Developing countries like India are facing grave traffic congestion and pollution problems in recent decades owing to increasing vehicular traffic and rapid urbanization. Affordable and accessible public transport services in the cities help decrease traffic congestion that occurs due to the movement of personalized vehicles on roads. Therefore, the accessibility to a sustainable public transport service (PTS) is vital for general masses, especially for low-income households in cities. Recently, general transit feed specification (GTFS) transit data is now increasingly used for analyzing the different types of transit performance parameters like transit availability within a specific time window, frequency, geographical coverage, etc., between different locations in a service area. Moreover, increased availability of GTFS-based transit data created new opportunities for the development of methodologies for investigating accessibility to PTS in the urban areas. In this paper, accessibility of bus transportation using GIS and (GTFS) dataset acquired from Delhi Transport Corporation (DTC) has been used for investigating accessibility to the key points of interest (POIs) in the National Capital Territory (NCT) of Delhi. In this analysis, several characteristics of bus services, travel time, frequency of transit between bus stops, accessibility of origins to destinations, mainly how easily passengers can access their destinations by public transportation have been analyzed. It can be concluded from the investigation that some regions in NCT require improvement in terms of accessibility of public transit facilities, and these conclusions can provide valuable results to the transport planners seeking solutions to provide access to public transportation.

G. H. Rezai · V. Singh (✉) · V. Nayak
Department of Civil Engineering, Motilal Nehru National Institute of Technology Allahabad, Prayagraj, Uttar Pradesh 211004, India
e-mail: varun@mnnit.ac.in

G. H. Rezai
e-mail: grezai@mnnit.ac.in

V. Nayak
e-mail: vibham@mnnit.ac.in

Keywords Accessibility of bus transport · Bus travel time · Dynamic accessibility · Bus service quality

1 Introduction

Metropolitan cities in emerging economies like India, Brazil, etc., have undergone rapid infrastructural development as well as economic growth in the last several decades on transportation facilities, infrastructures and land use in cities. As a result, urban regions are now characterized by extensive spatial and social isolation and unequal distribution of transport infrastructure in terms of urban mobility and accessibility, particularly for low-income households [1–4]. To address these spatial and social disparities, expansion of public transport services (PTS), viz. bus transportation and mass rapid transport system (MRTS) has been planned or implemented by public organizations and institutions in the developing countries. As an example, in Delhi, India, (which is a developing country), bus transportation and metro rails are now extensively utilized by masses for travel. To assess the effectiveness of these infrastructural expansions, accessibility studies pertaining to public transport services are gaining prominence. Hence, correctly measuring and monitoring the accessibility at different places in the urban area are essential for effective urban transportation planning [5]. Moreover, nowadays with the increased availability of general transit feed specifications (GTFS)-based transit data in the public domain, researchers have utilized them to investigate spatiotemporal accessibility to public transport services in a more comprehensive manner. In this paper, spatiotemporal accessibility between important locations for bus transport using the GTFS dataset and GIS-based analysis has been presented while taking NCT, Delhi, as a case application of a typical metropolitan city in a developing country. In this paper, the spatial aspect of accessibility is considered in terms of time required by the travelers to reach destinations from their origins using buses and temporal aspects of the accessibility investigated in terms of availability of buses between different points of interest during the different times of the day.

2 Literature Review

PTS in the cities has gained more attention due to its positive impacts on the traffic conditions as well as to expand the beneficial aspects of sustainable public transport and the living quality in urban areas. Many urban planners and transportation engineers emphasize that the primitive objective of transportation planning must change from mobility (e.g., speeding up or congestion alleviation) to accessibility [6]. Accessibility annotates the practicality of reaching desired locations within an acceptable travel time, and its definition emphasizes door-to-door access and accounts for the mode choice. Two types of accessibility have been investigated by researchers, viz. person accessibility and place accessibility [7–9]. Person accessibility describes places that are accessible to a person, and it depends on income, availability of mode of transport and time constraint of that specific person. Place accessibility describes characteristics of places and accessibility of those places to a person using different modes of transport. Place accessibility depends upon the type of locations and the transportation network connected to these places. In this paper, place accessibility is considered for analysis. Most of the literature regarding accessibility use travel time as the primary indicator because time is closer to a traveler's experience than distance [10].

Accessibility to PTS has been quantified in terms of spatial and temporal aspects. The spatial dimension of accessibility is quantified as the aggregated cost, viz. aggregation of walking time to or from transit stops and travel time by vehicle along the transit route. However, the spatial aspect is not sufficient to define accessibility; temporal availability of transit vehicles has a significant bearing on accessibility [11–14]. Temporal aspects comprise temporal variability of availability of public transit between different locations in the city within the given time interval [13].

Accessibility for PTS has been investigated independently from other modes of transport using different approaches. The LAUTAI indexing model developed by Yigitcanlar et al. [15] was used to measure levels of accessibility to essential community services, and it provides a grid-level origin-based accessibility index by measuring transit travel times and walking distances. The transit opportunity index (TOI), a method established by Mamun et al. [16], integrates measures of spatiotemporal accessibility with connectivity measures to capture various services and opportunities provided to transit passengers in an area.

Mavoa et al. [17] assert in their accessibility study that from each particular bus stop, approximately four number of trips happen hourly. They concluded that accessibility studies are essential for encouraging the mode shifts to reduce the dependence of travelers on the personalized modes of transport. The approach proposed by them is not appropriate, in the case when travelers use multiple transit systems for their trips with varying frequencies. Their approach is not suitable to determine the real rate of trips or transit frequency. Moreover, as asserted by the authors, mode shift from cars to PTS is not always relevant, as shown by Turcotte [18] in his analysis of commuters' travel in Toronto, Canada. Recently, reported works incorporated the equity cost, mode share relative to auto and public transit and developed some indexes for public transport accessibility measurements, including variability over time of day (temporal) transit services [11, 12, 14]. Kim and Lee [5] developed an extensive accessibility index based on GTFS data to study access to jobs that consider the number of other possible routes served by transit and the number of trips for PTS service. The travel time has been calculated as a combination of total travel time, including the travel time inside the vehicle and out-of-vehicle based on the shortest path method, which utilizes the algorithm developed by Dijkstra's. Lakhotia et al. [19] investigated the accessibility of people to bus stops in Delhi. They carried out the accessibility audits for 360 stops out of 3210 available bus stations across the whole city. They reported that the scores for pedestrian accessibility were considerably less for the surveyed bus stops in NCT. The authors' approach is based on the safety of pedestrians around bus stops rather than travel time or mobility of PTS.

Transit service has been measured independently by various transit agencies/organizations in different ways, among these, common measures include spatial coverage using Euclidean distance around bus stops (like 400 m buffer) [20], and actual walking distance [21]. TCRP Report [20] proposed the availability measures like transit frequency, capacity and service span [20] to assess the quality of PTS. Mamun [22] utilizes the service gap measures to identify where transit service is lacking.

Google API and GTFS published by different transit agencies are now frequently used by researchers to study accessibility. Farber et al. [23] used GTFS datasets to measure travel impedance of PTS from census blocks to the adjacent supermarkets considering different times of a selected day in Cincinnati, Ohio. They reported significant temporal variability in terms of travel times for a typical day. Bok and Kwon [13] suggested a methodology using the GTFS and population data to investigate the accessibility of public transport in the cities. Pereira et al. [24] assessed the accessibility alteration of public transit improvements caused by major events like the Football World Cup and the Olympic games.

As concluded in the above research works, assessment of accessibility to PTS in Delhi with the available of GTFS dataset for the city is not investigated comprehensively. The presented methodology complements the previous studies to address accessibility gaps about PTS in Delhi using the GTFS dataset.

2.1 Study Area

Delhi is the sprawling metropolitan, most populated and the fastest-growing city. The location of Delhi is from latitude $28.24°N$ to $28.53°N$ and longitude from $76.50°E$ to $77.20°E$, which is having an area of 1483 km^2 and having a shared administrative boundary that abuts the Indian states of Haryana and Uttar Pradesh. The abundant usage and upsurge of private vehicles in Delhi cause traffic congestions and environmental issues in the whole city. Only public transport systems are more efficient users of transportation infrastructures. Public transit replaces personalized vehicles and alleviates city-wide traffic congestions. It enhances road safety and decreases the negative impacts of traffic on the environment.

In particular, bus transport is an essential mode of transportation in the NCT of Delhi, as more than two-thirds of the population depend on it [25]. There are two crucial components of public transport in Delhi: bus and metro. Currently, Delhi metro ridership is 2.5 million, and it is expected to expand after the construction of additional corridors. However, daily average passenger ridership on the Delhi Transport Corporation, DTC and cluster buses is 4.3 million according to recent reports of the [26]. Therefore, the accessibility to bus transportation in NCT is essential because a large number of people rely on bus transit, and it is crucial for providing an integrated transportation system in addition to other modes of transport.

As reported in the Economic Survey of Delhi, 2018–19, total motorized vehicles in NCT of Delhi are around 11 million. According to the report, the percentage of buses is less in comparison with other modes of transport, but the efficiency of the bus as a mode of transportation is high.

Delhi Transport Corporation (DTC) is the leading PTS service provider of bus transport. Delhi is one of the biggest users of bus transport in the world, as reported by Chand and Chandra [27], in which buses operate based on CNG power. The fleet sizes of DTC buses reported by Delhi National Urban Transport Helpline [28] are 4237 buses that an average number of 3647 operate in a full day. The bus network that covers the entire Delhi is approximately 16,200 km in length in which the number of routes available, as mentioned by the transport department, is 657. The "cluster bus" operates by private sectors, which runs 1500 buses has launched in Delhi by 2011, in which numerous clusters work by the bidding method under the gross cost model. The routes of this scheme are divided into 17 clusters according to their areas of operation.

3 Methodology

Methodology for the assessment of the accessibility to bus transit services comprises different steps. As a first step, the GTFS data is acquired, preprocessed and structured as SQLite database. In the second step, the spatiotemporal network dataset (compatible for processing in ArcGIS software) was generated using open street road network and GTFS database. In the third step, the OD matrix was developed using the network dataset. In the final step, step-wise analysis for the measurement of accessibility has been carried out in three stages. First, bus transport travel times are assessed between sets of origins and destinations. Then, these travel times are converted to the measurement of accessibility, which directly impacts accessibility. Finally, the statistical analysis was performed to predict the behavior of public transit accessibility (Fig. 1).

4 Data

GTFS dataset was acquired from the open transit data web portal (https://opendata. iiitd.edu.in/). The dataset is published and updated by the Department of Transport (Government of NCT of Delhi) in association with IIIT-Delhi using street data extracted from OpenStreetMap (OSM). The network dataset has been created using network analyst extension, GTFS dataset and OpenStreetMap road network dataset for the spatiotemporal analyses. The following maps in Fig. 2 represent (a) our study area and the 3210 bus stop locations and (b) origin and destination maps and matrix.

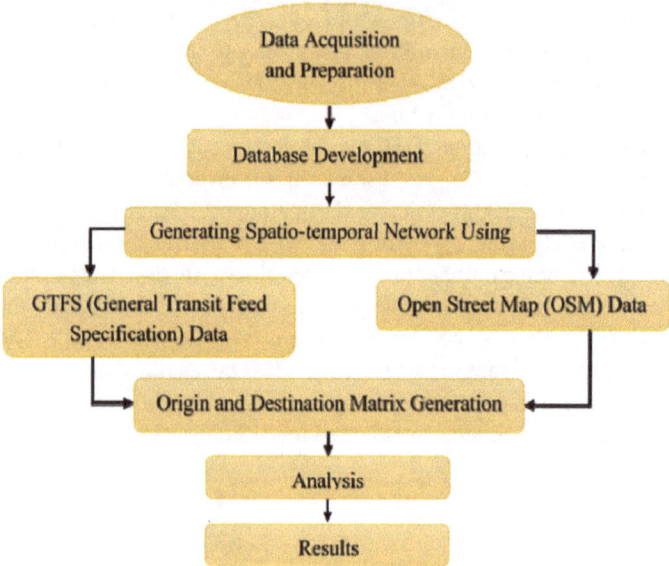

Fig. 1 Flowsheet of detailed methodology

5 Travel Time Calculation

The spatiotemporal network has been created using Add GTFS to a network dataset tool developed by ESRI, GTFS dataset and OpenStreetMap road network data. The tool and datasets are used to create a complete road network for transit travel time estimation. The bus travel time for each set of origin destination (OD) pairs has been calculated using the OD cost matrix solver functionality of the network analyst extension of ArcGIS software. The solver tool uses a customized algorithm based on Dijkstra's algorithm to calculate OD cost matrices in terms of travel time. Moreover, an ingress, egress and transition walking speed of 5 km per hour have been assumed in the analysis. The delay of transit is 0.4 min for the transfer from bus lines (boarding) to bus public transport and 0.3 min for changing from transit facility to roads (alighting), which were taken into consideration for analysis. A pedestrian restriction has been considered for the analysis for restricting travelers to avoid walking on major roads. This accessibility restriction is considered to prevent the interruptions of the traffic flow due to the users of PTS.

For the presented work, the journey time is the time taken by the travelers to travel from a given origin to destination using the computed optimal path in the multimodal network at that particular time of the day. For calculating the bus travel time along the path of the transit facility, the network utilizes transit evaluator of "add GTFS to a network dataset" tool. The transit evaluator finds the travel impedance across the bus lines by searching the transit trips available in the GTFS schedules, then summing both waiting and riding time from the present stop to the next one. For example, a

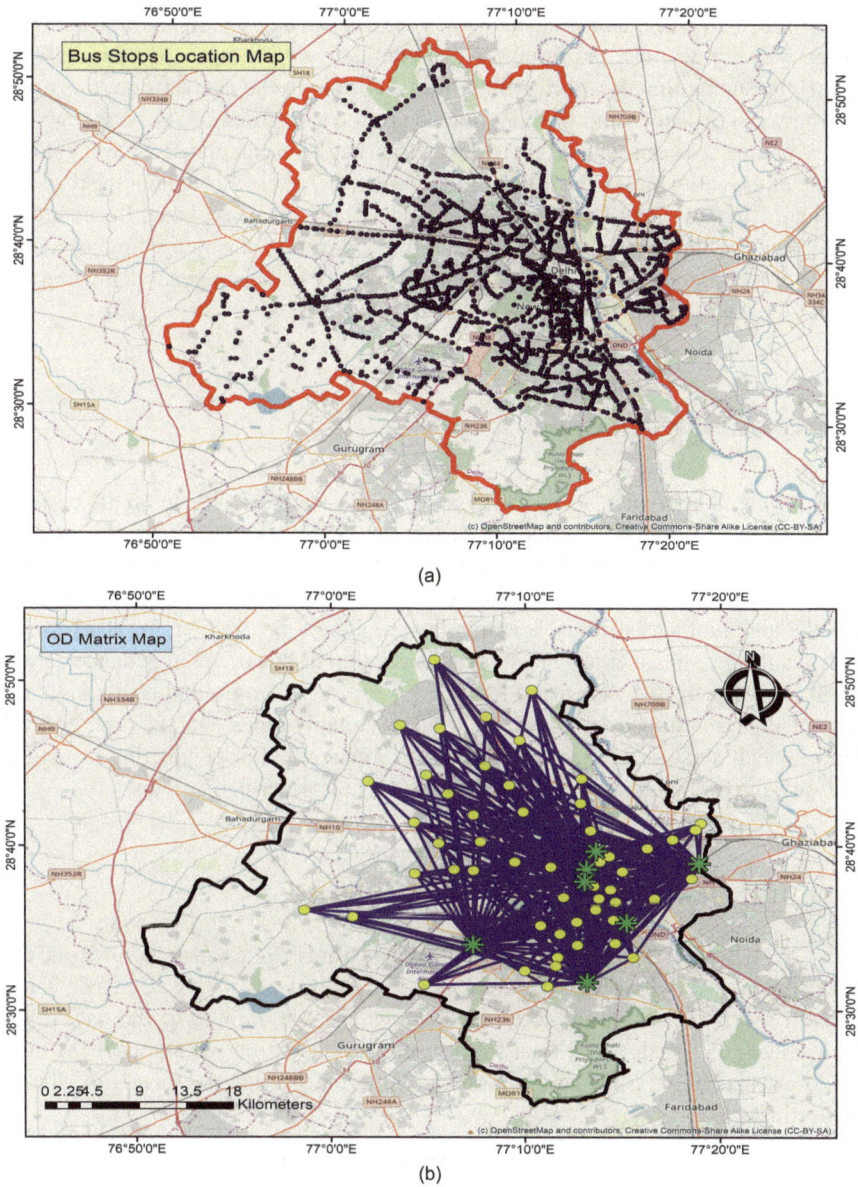

Fig. 2 **a** Study area NCT of Delhi and bus stop locations, **b** origins and destinations maps and matrix

travel time of 20 min is a time calculated on the basis of aggregation of time incurred during walking, waiting and riding, which cumulatively becomes 20 min. Walking time is the summation of walking time from an origin to the nearest stop, transition time between two stops and walking time from stop to destination. For our analysis, the travel time between each origin to destination has been calculated for every single minute of the entire day as the travel time between OD alters throughout the day. For the spatiotemporal analysis, Tuesday has selected a typical weekday. Because OD travel time from each origin to all destination pairs varies during the time of a day, the bus travel time matrix for each minute of the day has been generated using the travel time statistics tool.

6 Accessibility Analysis

In the proposed approach, accessibility analysis has been carried out on the basis of the computation of bus travel time for measuring the variability of the transit facility. The quantitative assessment of relating variability into accessibility during the whole day, specific measuring of accessibility for each time of day, has been used to analyze how the temporal accessibility is within the specified time. At first, for all origin to each destination, the average and standard deviation of bus travel impedance across the time period from 6 AM to 10 PM for a given set of origin and destination are computed. It is assumed that this time window suitably represents the time interval during which most of the transit trips happen frequently. For conducting statistical analyses, the following formula has been used that is described below:

$$\mu_o^d = \frac{\sum_t To, t, d}{M} \tag{1}$$

where $T_{o, t, d}$ is the estimated bus travel impedance from origin o, at departure time t to the destinations d (measured in minutes). For the analysis, $T_{o, t, d}$ has been calculated form all seven origins, to all sixty destinations at every minute of the day. μ_o^d is the average bus transit travel time from origin to destination within specified time window of size M. σ_o^d is the standard deviation for the average bus travel time for all the destinations.

$$\sigma_o^d = \sqrt{\frac{\sum t (To, t, d - \mu_o^d)^2}{M - 1}} \tag{2}$$

μ_o^d and σ_o^d are used in measuring mean levels of access to destinations and variations of mean travel time. Accessibility of an origin can be assessed using these two factors as proposed by Farber et al. [23]. Table 2 depicts how these two factors lead to different accessibility results.

Table 2 Statistical markers for quantitative assessment of the accessibility [23]	Low travel time μ	High travel time μ
Low standard deviation σ	Best	Worst
High standard deviation σ	Satisfactory	Unsatisfactory

The best accessibility condition is for an origin that has small mean travel time and low standard deviation, and it is categorized as well connected by bus throughout the entire day. This best condition means that the people have shorter travel times with less variability to destinations during the day. Conversely, high average travel time with low standard deviation means that an origin is not served well and is poorly connected to the destinations in terms of travel time and spatial availability. Less average travel time with high standard deviation depicts the most reliable and robust accessibility with a slight variation of the travel times from their mean value. In contrast, a high mean and standard deviation values indicate that the transit availability has reduced and has poor accessibility with occasional peaks.

Moreover, the distribution of public transport travel times considering various time windows is investigated. This distribution of travel time elaborates on the variability of bus facilities. As an example, if the distribution of travel time is uniform, it indicates no variability. On the other hand, if travel time varies frequently, then it indicates that the changes in travel time are significant. The variability of bus travel time is quantified using the coefficient of variation. A coefficient of variation (CV) indicates the variation as a percentage of the mean or SDT divided by mean.

$$CV = \frac{STD}{TT} \times 100 \tag{3}$$

where CV = coefficient of variation (%), STD = standard deviation of travel time, TT = mean of travel times, CV has been chosen as a meaningful comparison between two or more values of variations. Because they are quantified in the form of a percentage, the variability from different sources is comparable even if they have various means or scales of measurements.

7 Results

Accessibility analysis has been carried out using the parameters described in the previous section. Results are visualized using maps and scatterplots depicting the variation of daily average bus travel time to all sets of destinations that were generated on the basis of the statistical analysis. Specific origin and destinations are considered for the analysis. The variation of T o, t, d, versus time of day is depicted in Fig. 3. Moreover, it is evident from Fig. 3 that travel times vary from 108 to 332 min, based

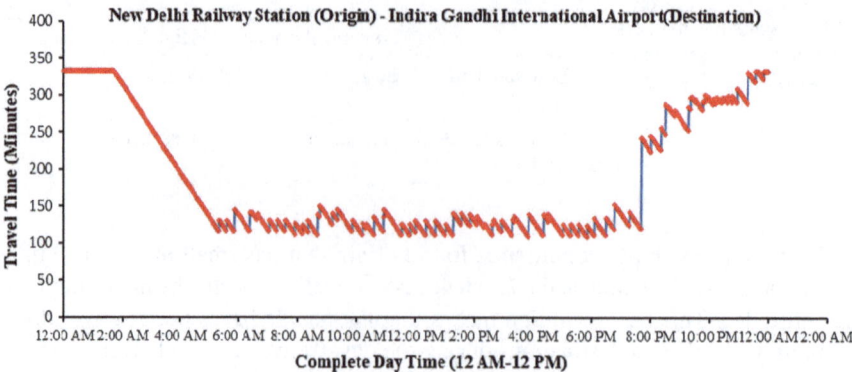

Fig. 3 Bus travel time variation on specified selected origin and destination during a complete day (24 h)

on the availability and timing of bus facilities. The travel time variations between the specified origin and destination are significantly higher depending on the time during a day and travelers prefer to avail other modes of transport like mass rapid transport system (MRTS) or taxi in this case. However, those people who are residents of these areas still use the bus service. Therefore, there is an urgent need to improve this mode of public transportation. This variability in bus transport travel times is the proportion to the variation of access to transit services directly.

The BetterBusBuffers tool developed by Morang [29] is used for generating the trips/hour map as shown in Fig. 4. The distance buffers created around bus stops are showing the average number of bus trips per hour accessible at different locations within the time window of 6 AM and 10 PM. The destinations of trips are not considered in this map, and it depicts only the frequency of bus facilities in NCT of Delhi between bus stops. The GTFS and OSM networks are used to calculate the average trips per hour available from stop to stop based on a 10-min walk buffer and color code for better visualization and comprehension.

Figure 5 depicts travel times from origins to the destinations in terms of the average bus travel time and standard deviation of travel impedance. The maps show that closeness to destinations results in quicker travel times, and significantly smaller standard deviations are also found in those places with shorter travel times. From Fig. 5, it is evident that areas are not spatially adjacent to destinations but have lesser travel times due to availability of a greater number of transit trips and network connectivity. In Fig. 5a, one example of the critical location is Indira Gandhi International Airport, which explicitly depicts less access by public transit from all seven selected origins, and the variations of travel impedance are greater in comparison with other destinations.

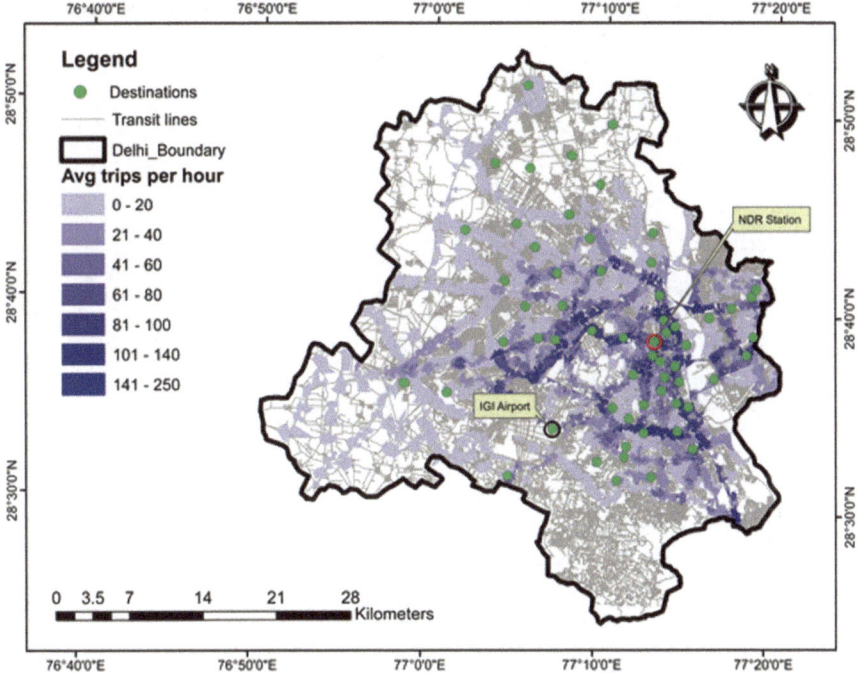

Fig. 4 Public transit frequency of bus service in NCT of Delhi

8 Conclusions

In this study, transit-based accessibility to destinations has been investigated for the NCT of Delhi. First, the degree to which the schedules of available trips of bus transport systems alter during the time of the day has been analyzed. It can be posited that accessibility from the origin to destinations not only depends upon the distance but also on the availability of the bus service during the particular time of the day between them. It has been observed that owing to the high frequency of trips from origins to remotely located destinations, the travel time using the bus is lesser. It can be concluded that access to destinations using buses exhibits a high degree of variability depending on the different time of a particular day and frequency of the bus service. Furthermore, this variability of bus transport relies on factors such as distance from the selected sets of origins to destinations and the presence of transport options between locations. Finally, in these specific case study findings, the continuous accessibility proposed analysis can be utilized to investigate the spatiotemporal variability in accessing of origin to the destination. Based on the availability of bus transit facilities, fewer people utilize buses and achieve consistent access to the destination during the day.

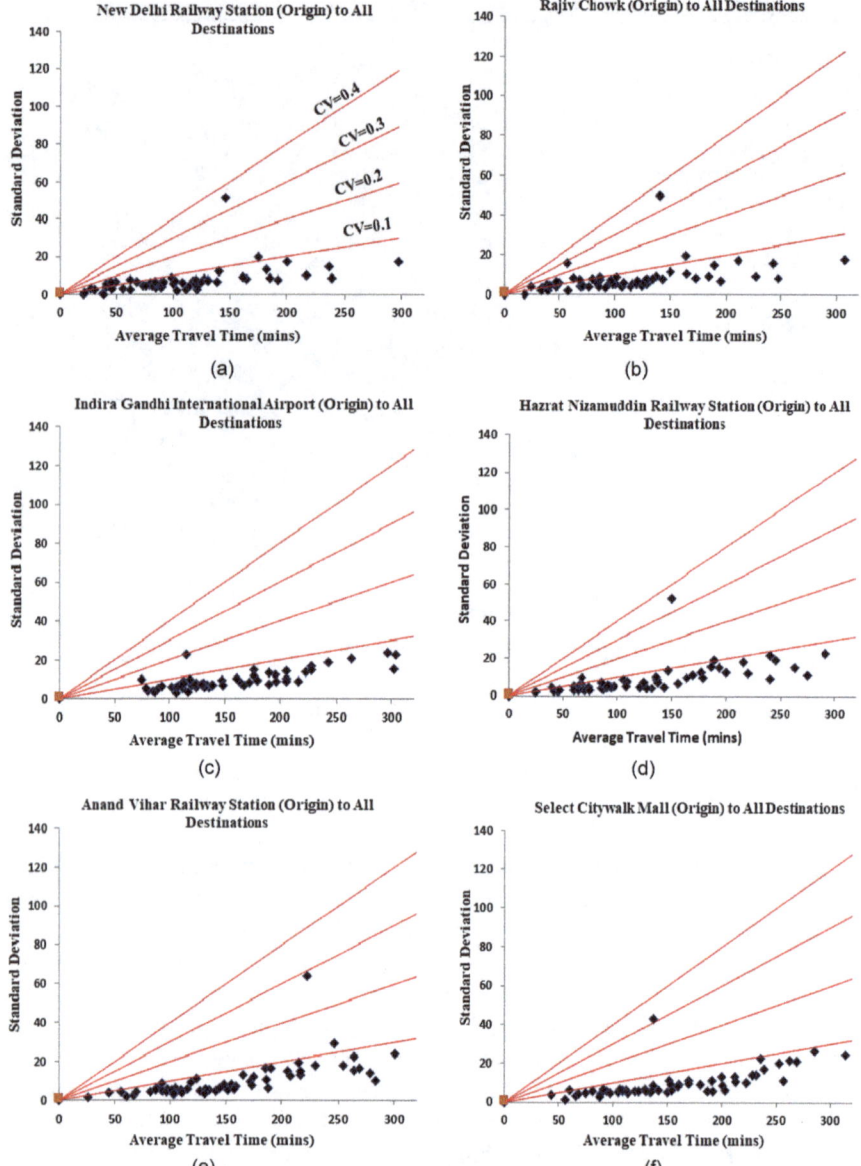

Fig. 5 Scatterplots of bus average travel time from seven important origins to all sixty destinations

Fig. 5 (continued)

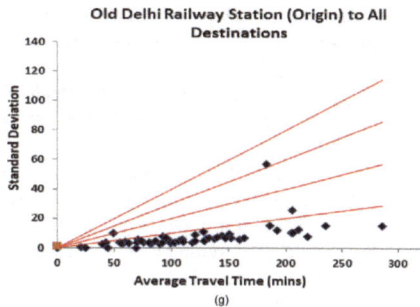

(g)

References

1. E.C. Delmelle, I. Casas, Evaluating the spatial equity of bus rapid transit-based accessibility patterns in a developing country: the case of Cali, Colombia. Transp. Policy **20**, 36 (2012). https://doi.org/10.1016/j.tranpol.2011.12.001
2. J.P. Bocarejo S, D.R. Oviedo H, Transport accessibility and social inequities: a tool for identification of mobility needs and evaluation of transport investments J. Transp. Geogr. **24**, 142–154. https://doi.org/10.1016/j.jtrangeo.2011.12.004
3. D. Hernandez, C. Rossel, Inequality and access to social services in Latin America: space–time constraints of child health checkups and prenatal care in Montevideo. J. Transp. Geogr. **44**, 24–32 (2015). https://doi.org/10.1016/j.jtrangeo.2015.02.007
4. D. Hernandez, Uneven mobilities, uneven opportunities: Social distribution of public transport accessibility to jobs and education in Montevideo. J. Transp. Geogr. **67**, 119–125 (2018). https://doi.org/10.1016/j.jtrangeo.2017.08.017
5. J. Kim, B. Lee, More than travel time: new accessibility index capturing the connectivity of transit services. J. Transp. Geogr. **78**, 8–18 (2019). https://doi.org/10.1016/j.jtrangeo.2019.05.008
6. J. Levine, J. Grengs, Q. Shen, Q. Shen, Does accessibility require density or speed? A comparison of fast versus close in getting where you want to go in US metropolitan regions. J. Am. Plann. Assoc. **78**(2), 157–172 (2012). https://doi.org/10.1080/01944363.2012.677119
7. M.P. Kwan, Gender and individual access to urban opportunities: a study using space-time measures. Prof. Geogr. **51**(2), 210–227 (1999). https://doi.org/10.1111/0033-0124.00158
8. H. Miller, Place-based versus people-based geographic information science. Geogr. Compass **1**(3), 503–535 (2007). https://doi.org/10.1111/j.1749-8198.2007.00025.x
9. K. Martens, Justice in transport as justice in accessibility: applying Walzer's 'Spheres of Justice' to the transport sector. Transportation **39**(6), 1035–1053 (2012). https://doi.org/10.1007/s11116-012-9388-7
10. A. Lovett, R. Haynes, G. Sünnenberg, S. Gale, Car travel time and accessibility by bus to general practitioner services: a study using patient registers and GIS. Soc. Sci. Med. **55**(1), 97–111 (2002). https://doi.org/10.1016/S0277-9536(01)00212-X
11. A. Owen, D.M. Levinson, Modeling the commute mode share of transit using continuous accessibility to jobs. Transp. Res. Part A Policy Pract. **74**, 110–122 (2015). https://doi.org/10.1016/j.tra.2015.02.002
12. S.K. Fayyaz, X.C. Liu, R.J. Porter, Dynamic transit accessibility and transit gap causality analysis. J. Transp. Geogr. **59**, 27–39 (2017). https://doi.org/10.1016/j.jtrangeo.2017.01.006
13. J. Bok, Y. Kwon, Comparable measures of accessibility to public transport using the general transit feed specification. Sustainability **8**(3), 224 (2016). https://doi.org/10.3390/su8030224
14. G. Boisjoly, A. El-Geneidy, Daily fluctuations in transit and job availability: a comparative assessment of time-sensitive accessibility measures. J. Transp. Geogr. **52**, 73–81 (2016). https://doi.org/10.1016/j.jtrangeo.2016.03.004

15. T. Yigitcanlar, N. Sipe, R. Evans, M. Pitot, A GIS-based land use and public transport accessibility indexing model. Aust. Planner **44**(3), 30–37 (2007). https://doi.org/10.1080/07293682.2007.9982586

16. S.A. Mamun, N.E. Lownes, J.P. Osleeb, K. Bertolaccini, A method to define public transit opportunity space. J. Transp. Geogr. **28**, 144–154 (2013). https://doi.org/10.1016/j.jtrangeo.2012.12.007

17. S. Mavoa, K. Witten, T. McCreanor, D. O'sullivan, GIS based destination accessibility via public transit and walking in Auckland New Zealand. J. Transp. Geogr. **20**(1), 15–22https://doi.org/10.1016/j.jtrangeo.2011.10.001

18. M. Turcotte, Commuting to work: results of the 2010 general social survey. Can. Soc. Trends **92**(August), pp. 25–36 (2011). https://www150.statcan.gc.ca/n1/pub/11-008-x/2011002/article/11531-eng.pdf

19. S. Lakhotia, K.R. Rao, G. Tiwari, Accessibility of bus stops for Pedestrians in Delhi. J. Urban Plann. Dev. **145**(4), 05019015 (2019). https://doi.org/10.1061/(ASCE)UP.1943-5444.0000525

20. Transit Cooperative Highway Research Program (TCRP) Report 165: Transit Capacity and Quality of Service Manual, 3rd edn. (Transportation Research Board of the National Academies, Washington DC, 2014). https://www.trb.org/Main/Blurbs/169437.aspx

21. P. Ryus, J. Ausman, D. Teaf, M. Cooper, M. Knoblauch, Development of Florida's transit level-of-service indicator. Transp. Res. Rec. **1731**(1), 123–129 (2000). https://doi.org/10.3141/1731-15

22. S. Mamun, Public transit accessibility and need indices: approaches for measuring service gap. https://doi.org/10.3141/2217-19

23. S. Farber, M.Z. Morang, M.J. Widener, Temporal variability in transit-based accessibility to supermarkets. Appl. Geogr. **53**, 149–159 (2014). https://doi.org/10.1016/j.apgeog.2014.06.012

24. R.H. Pereira, D. Banister, T. Schwanen, N. Wessel, Distributional effects of transport policies on inequalities in access to opportunities in Rio de Janeiro. Available at SSRN 3040844 (2017). https://doi.org/10.31235/osf.io/cghx2

25. T. Bhatia, M. Jain, Bus Transport in Delhi (No. 210). Working Paper (2009). https://ccs.in/internship_papers/2009/bus-transport-in-delhi-210.pdf

26. Economic Survey of Delhi, 2018–19. https://delhiplanning.nic.in/content/economic-survey-delhi-2018-19

27. S. Chand, S. Chandra, Improper stopping of buses at curbside bus stops: reasons and implications. Transp. Dev. Econ. **3**(1), 5 (2017). https://doi.org/10.1007/s40890-017-0033-1

28. Delhi National Urban Transport Helpline (Nov 2016). https://www.sutpindia.com/skin/pdf/DelhiNUTH.pdf

29. M. Morang, BetterBusBuffer tool, environmental systems research institute, https://www.arcgis.com/home/item.html?id=42e57c5ff9a0497f831f4fced087b9b0

30. R. Cervero, T. Rood, B. Appleyard, Tracking accessibility: employment and housing opportunities in the San Francisco Bay Area. Environ. Plann. A **31**(7), 1259–1278 (1999). https://doi.org/10.1068/a311259

31. M.Q. Dalvi, K.M. Martin, The measurement of accessibility: some preliminary results. Transportation **5**(1), 17–42 (1976). https://doi.org/10.1007/BF00165245

32. Delhi Transport Corporation: https://www.dtc.nic.in/home/delhi-transport-corporation-dtc

33. R. Goel, G. Tiwari, Access-egress and other travel characteristics of metro users in Delhi and its satellite cities. IATSS Res. **39**(2), 164–172 (2016). https://doi.org/10.1016/j.iatssr.2015.10.001

34. K.T. Geurs, B. Van Wee, Accessibility evaluation of land-use and transport strategies: review and research directions. J. Transp. Geogr. **12**(2), 127–140 (2004). https://doi.org/10.1016/j.jtrangeo.2003.10.005

35. S.L. Handy, D.A. Niemeier, Measuring accessibility: an exploration of issues and alternatives. Environ. Plann. A **29**(7), 1175–1194 (1997). https://doi.org/10.1068/a291175

36. W.G. Hansen, How accessibility shapes land use. J. Am. Inst. Plann. **25**(2), 73–76 (1959). https://doi.org/10.1080/01944365908978307

37. A. Owen, D. Levinson, Modeling the commute mode share of transit using continuous accessibility to jobs, in 93rd Annual Meeting of the Transportation Research Board. Washington, DC (2014). https://doi.org/10.1016/j.tra.2015.02.002

38. A. Páez, D.M. Scott, C. Morency, Measuring accessibility: positive and normative implementations of various accessibility indicators. J. Transp. Geogr. **25**, 141–153 (2012). https://doi.org/10.1016/j.jtrangeo.2012.03.016

Extraction of Information from Hyperspectral Imaging Using Deep Learning

Anasua Banerjee, Satyajit Swain, Mainak Bandyopadhyay, and Minakhi Rout

Abstract Keeping pace with the rapid change of time, very powerful deep learning techniques have become available. There are many optical analysis techniques that are used to identify different objects from large-scale images. Hyperspectral is one of such techniques, which is used mainly when the images from the satellite are captured and are used to identify different objects. Hyperspectral images are made up of a large number of bands naturally. Thus, extracting information from the satellite comes with many problems and challenges. But with the use of powerful deep learning (DL) methods, the earth's surface can be precisely explored and analysed. The combination of spatial and spectral information helps to track and find out remotely sensed scrutinized data everywhere. However, the presence of high-dimensional features along with various bands hinders the accuracy in hyperspectral imaging (HSI) analysis. Thus, a large number of bands and dimensions in hyperspectral images must be reduced using some linear or nonlinear dimensionality reduction (DR) techniques. By eliminating all these redundant information, data can be classified and interpreted in a far better way. Mainly three hyperspectral datasets have been extensively used for this purpose, namely Pavia University, Indian Pines, and Salinas Valley. In this paper, the main focus is on the use of principal component analysis (PCA), a technique for DR, followed by that the use of a deep neural network, such as 3D convolutional neural network (CNN) for classification of the hyperspectral datasets. Subsequently, all these classified datasets are then compared with the respective

A. Banerjee (✉) · S. Swain · M. Bandyopadhyay · M. Rout
School of Computer Engineering, Kalinga Institute of Industrial Technology, Bhubaneswar, Odisha, India
e-mail: anasua123.banerjee@gmail.com

S. Swain
e-mail: swain.satyajit2011@gmail.com

M. Bandyopadhyay
e-mail: mainak.bandyopadhyayfcs@kiit.ac.in

M. Rout
e-mail: minakhi.routfcs@kiit.ac.in

© The Author(s), under exclusive license to Springer Nature Singapore Pte Ltd. 2021
M. Bandyopadhyay et al. (eds.), *Machine Learning Approaches for Urban Computing*,
Studies in Computational Intelligence 968,
https://doi.org/10.1007/978-981-16-0935-0_3

ground truth images, analysing the performance of the DL model based on various metrics.

Keywords Deep learning (DL) · Hyperspectral imaging (HSI) · Dimensionality reduction (DR) · Principal component analysis (PCA) · Convolutional neural network (CNN)

1   Introduction

The term hyperspectral refers to the measuring o f a very large number of wavelength bands, containing a huge amount of spectral information. All this information is used to detect and recognize the shape and size of different ground cover materials. Many optical techniques are used to identify different objects from a huge amount of image datasets. Among these optical techniques, hyperspectral imaging is broadly used for classification and analysis of earth's observation for many potential applications. Within the range of human eye vision, only three different types of bands are present, i.e. red, green, and blue; whereas, in hyperspectral images, there exist several hundred different bands. All these channels exist in the form of a spectrum of different ranges, chemical structure, and texture. Consequently, due to the effect of sunlight, different objects absorb and reflect the different amounts of light, and thus they appear to be different from the visible human eye. Different colours signify distinct properties, such as the green colour signifies the presence of healthy vegetation, whereas the other pale colours reflect the presence of unhealthy, pest infected, and poorly developed vegetation in various areas of the earth's surface. Through hyperspectral imaging analysis, minute monitoring, tracking, focusing, and subsequently analysing the imminent happenings can be obtained precisely in various fields of study and investigation like remote sensing, environmental monitoring, agriculture precision, and food analysis as shown in Fig. 1 [1, 2]. Many developments are being made on hyperspectral imaging, and these protracted efforts are being conducted, as a result of which new tools are evolving gradually. With the help of all these tools, various environmental changes are being tracked. New sensor technologies are capable of covering large and vast surfaces of the earth, as a result of which all these technologies are used in remote sensing applications to explore hidden information of satellite images.

The presence of rich spectral and spatial information and efficient techniques makes hyperspectral imaging analysis prominent in different areas like safety evaluation, food quality, forensic document scrutiny, and homeland safety ensuring. The viability of a seed can be detected after analysing different seeds through a hyperspectral image when the reflectance spectrum is plotted. The precise analysis of the different spectral features is impossible to be observed through the naked eyes. It is also much useful when applied in the field of medical research, for example, cell and wound analysis, detection of the early stage of cancer, retinal diseases, etc. The presence of fine spectral resolution in these images is highly useful when applied in the areas of forensic laboratory exploration and investigation, for example, fingerprint reading, questioned document scrutiny, gun powder residue study, and bloodstain visualization. It is also used in the study and analysis in the food sector, for example, finding the freshness of the fish and the presence of sugar in melons, etc.

With the presence of a large number of spectral channels, hyperspectral images pose high-dimensional features. To avoid any redundant information that increases the cost of operation, or hinders the accuracy and performance, the large dimensions of the hyperspectral image must be reduced, preserving some meaningful properties of original data. Hence, some dimensionality reduction techniques can be used for this purpose. Further, if the band selection method is applied, then the prevention of redundant information can be expedited, and also the cost complexity is also reduced, providing better accuracy to the deep learning models. Here three datasets, namely Pavia University, Indian Pines, and Salinas have been used for our purpose. With a view to extracting spatial information, the principal component analysis (PCA) technique is used side by side, which in turn reduces the high-dimensional nature of these datasets. Subsequently, 3D convolutional neural network (CNN) has been applied on these datasets, aiming at encoding spatial and spectral information for classification purposes.

It may be presumed that this chapter "Extraction of Information from Hyperspectral Imaging using Deep Learning" is relevant to what is being propounded through the main book "Machine Learning in Urban Computing". As detailed information provided by satellite image is always very much appreciated everywhere, this appreciable information relates to tracking and monitoring changes of land cover, an environment in urban areas. If the quality of satellite resolution is very high, it becomes very much useful to monitor the health of vegetation, and it also gives detailed information about environmental condition analysis for urban planning and development programs. Hence, this chapter may enrich the contents of this book.

2   Hyperspectral Imaging

Any form of change on the earth's surface, regarding vegetation, water level or soil erosion is detected by many narrow contiguous spectral bands present in hyperspectral images [4, 5]. Hyperspectral imaging typically includes 63 bands, 27 of which are in visible form and near infrared (ranging from 0.4 to 1.0 μm). Again, one short wave infrared (ranging from 1.0 to 1.6 μm) and 28 other shortwave infrared bands (ranging from 2.0 to 2.5 μm) are used to depict minerals, and rest six are thermal bands. Hyperspectral imaging is composed of 100–200 spectral bands with a narrow bandwidth (ranging from 5 to 10 nm), whereas multispectral imaging is composed of 5–10 bands with relatively large bandwidth as shown in Fig. 2 [6, 7].

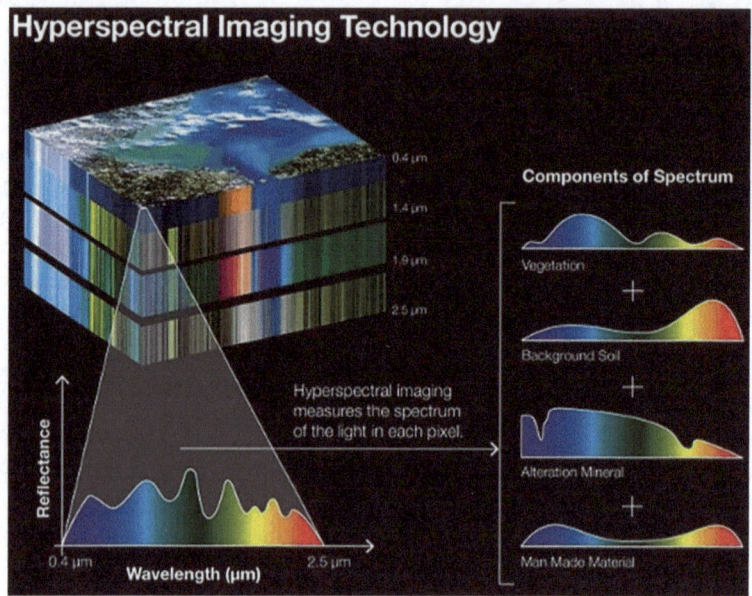

Fig. 1 Hyperspectral imaging technology [3]

2.1 Spatial–Spectral Classifications

The spatial resolution of a sensor is interpreted by taking one pixel as the ground area image. If the spatial resolution is high, every pixel will represent a small square area of the ground. As a result of which, the identification of the features of the image can be done in a better way [8, 9].

When an image is captured via the satellite mode, then the spatial resolution functions let us know how much area is covered, as well as how much detailed information can be obtained from that image. The spatial–spectral dimension of an image is given by $A \times K \times S$, where A, K, and S represent the width of the image, height of the image, and the number of bands in an image, respectively. For example, in an image of size $256 \times 256 \times 32$, 256 denote the width and height of the image, and 32 represents the total number of bands in the image, respectively [5]. The flowchart for integrated HSI analysis is shown in Fig. 3.

The spectral information received through a satellite with all its data is stored in separate bands. The spectral bands are of different types like red, green, blue, NIR, MID-IR, thermal, etc. The bands ranging from 0.76 to 0.90 μm help us to track and identify vegetation, crop, and land water level. Similarly, the bands ranging from 1.55 to 1.75 μm help us to detect the proportion of moisture contained in a plant. The bands which range from 2.08 to 2.35 μm measure the amount of radiation that comes out from the surface of the earth. Further, the bands ranging from 2.08 to 2.35 μm help us detect geological rock formation. Among the three primary bands,

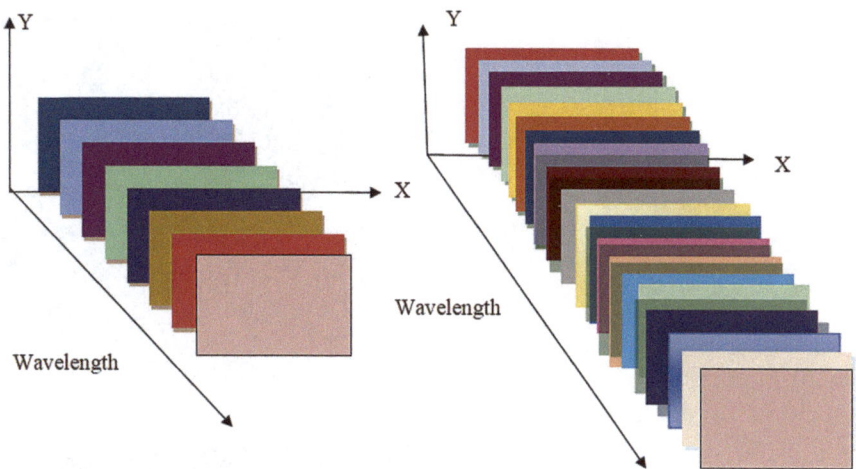

Fig. 2 Multispectral (left) and hyperspectral (right) bands

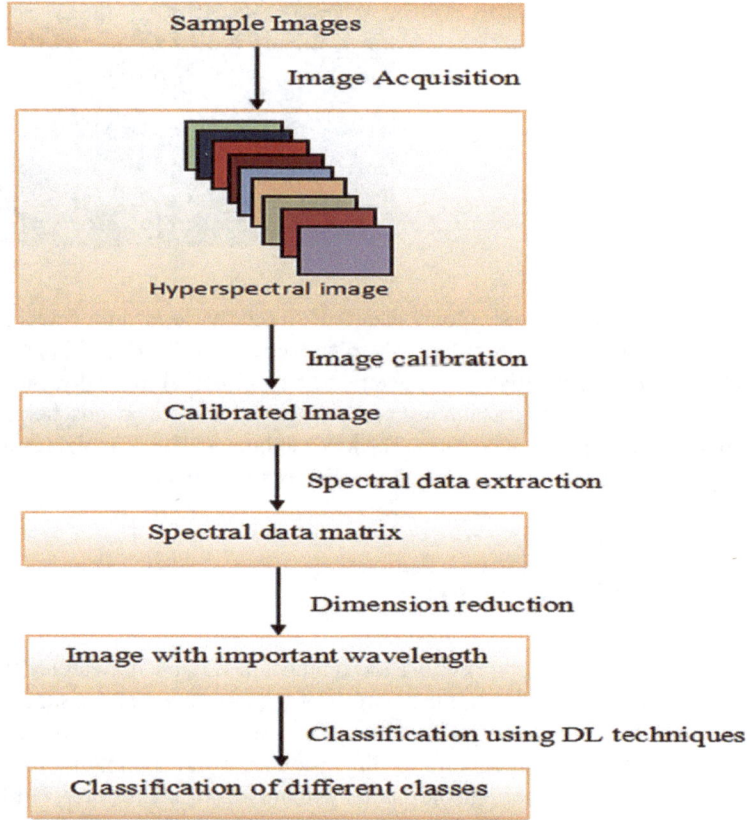

Fig. 3 Flowchart for integrated HSI analysis

Fig. 4 Low spatial
resolution

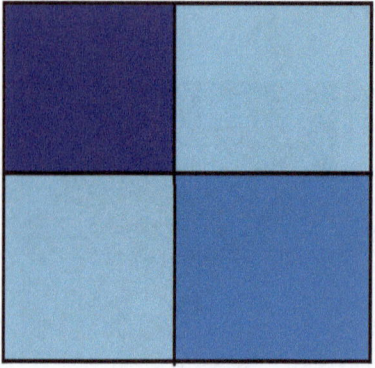

Fig. 5 Medium spatial
resolution

i.e. red, green, and blue, the red colour band (0.63–0.69 μm) is used for detecting the
discrimination of vegetation. Similarly, the blue colour band (0.45–0.52 μm) helps
us in detecting the characteristic of vegetation and soil use. And finally, the green
colour band (0.52–0.60 μm), which lies between the red and blue bands, helps us to
explore the healthy vegetation on earth. The low, medium and high spatial resolution
of an image are depicted in Fig. 4, Fig. 5 and Fig. 6, respectively.

3   Dimensionality Reduction

The dimension of a dataset comprises of a total number of variables or features present
in it. We should bear in mind that if the number of features or variables is increased,
the model may not be able to predict the classification accurately. To overcome
this problem, many linear and nonlinear dimension reduction techniques exist that
preserve the information present in the dataset without losing any features. Further, it
helps to avoid overfitting problems, thereby reducing the cost of complexity. Dimension reduction techniques are useful in matters like visualization, reduction in time

Fig. 6 High spatial
resolution

and space complexity, feature extraction, anomaly detection, and reduction in noise
from the dataset [10]. The steps to perform PCA and Isomap are depicted in Fig. 7
and Fig. 8 respectively. Hyperspectral imaging consists of both spatial and spectral
information with a large number of bands. Thus, applying dimension reduction techniques prior to classification will help to predict the different classes present in the
dataset accurately and precisely [11]. The classification of different classes using
hyperspectral image is shown in Fig. 9.

4 HSI Classification

Neural network is a system of embedded artificial neural networks interconnected
with each other, as a result of which all information messages get transferred from one
neuron to another neuron. Individually, all neurons have their own weights and bias,
which are changed in their training processes. Consequently, the presented model
can efficiently recognize the pattern of the features in the images in a precise way.

Convolutional neural network is a sector of neural network that can recognize,
classify, and identify the face of any object accurately. In CNN, there are many
layers like the input layer, hidden layer, and output layer, where each layer helps
to extract the features of an image. Previously, this extraction was done by hand-
designed method, but this was not as efficient as CNN. The kernel which is present
in a convolutional neural network need not work with the whole image itself, and
instead, it works on small patches at one time. Consequently, upon working on the
patches one after the other, we get a clear and precise result form on the whole image,
including the detection of the edges in a composite way.

Fig. 7 Steps to perform
PCA

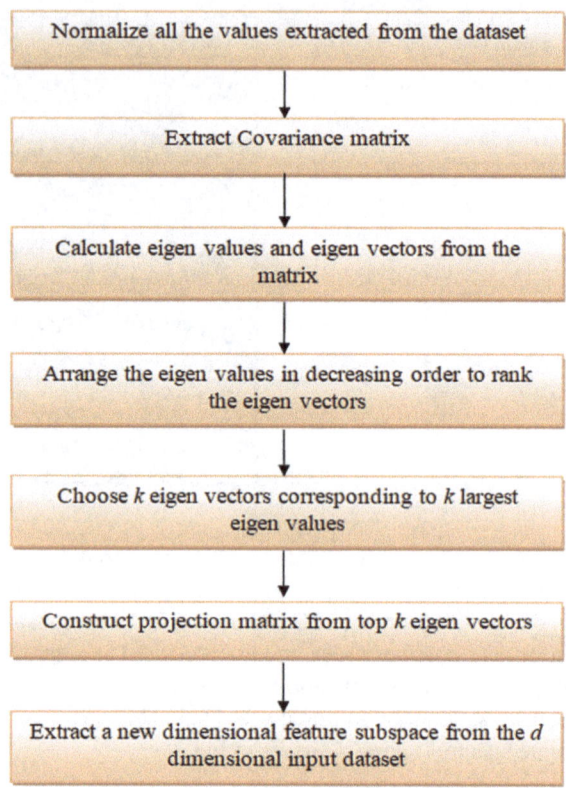

Normalize all the values extracted from the dataset

Extract Covariance matrix

Calculate eigen values and eigen vectors from the matrix

Arrange the eigen values in decreasing order to rank the eigen vectors

Choose k eigen vectors corresponding to k largest eigen values

Construct projection matrix from top k eigen vectors

Extract a new dimensional feature subspace from the d dimensional input dataset

4.1 Deep Learning Overview

Deep learning and machine learning are two integral parts of artificial intelligence. DL can help in detecting complicated problems with the help of hidden layers, which are typically present between the input and output layers [13]. Convolution neural networks have convolution filters, pooling operators (max-pooling, average pooling, global average pooling, etc.), and fully connected layers as its constituents. The model is made to run executing these layers in a step-by-step manner as follows (Fig. 10). At first, the original input image is made to pass through the convolution layer. Here a filter of size, say 3×3, is used, where the mathematical calculations and operations are carried out over the image according to the convolution filter. All these functions result in producing a feature map for each filter with the help of an activation function. For this purpose, the max-pooling operator is generally used, given by the function max[0, x], to draw out the maximum value revealed by the filter sliding over the image. Followed by that, the dense layer, with its own weight and bias, collects input from the preceding layers with all their activation values. Finally, the fully connected layer is used that does the function of flattening the output after connecting all the neurons of all the preceding layers. The flatten layer has the most important function

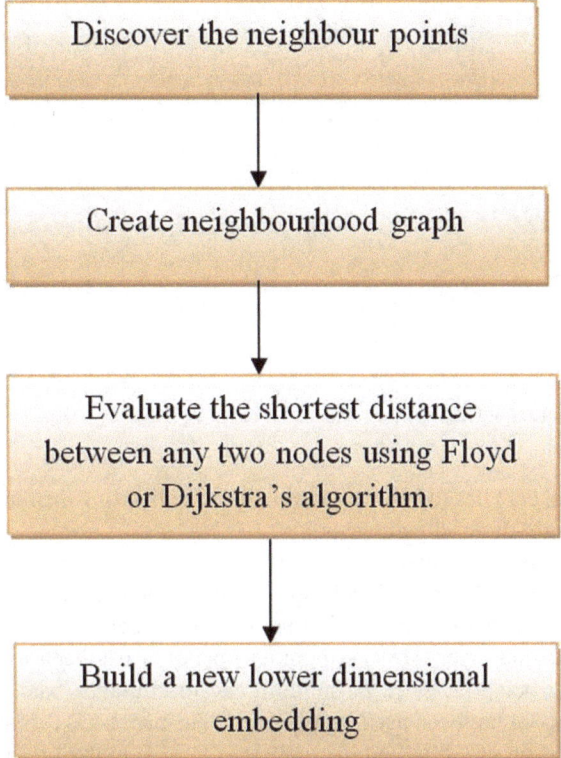

Fig. 8 Steps to perform Isomap technique [12]

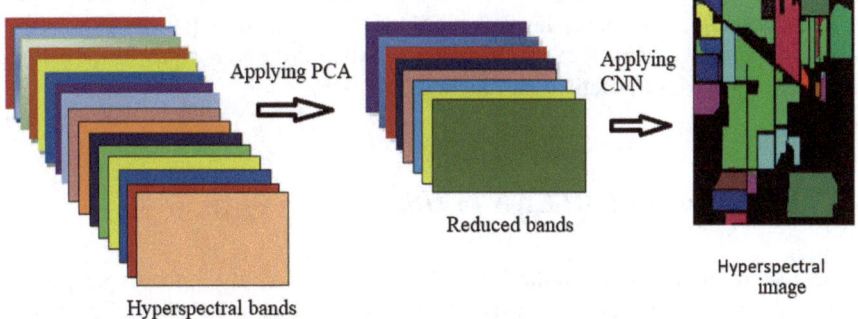

Fig. 9 Classification of different classes using hyperspectral image

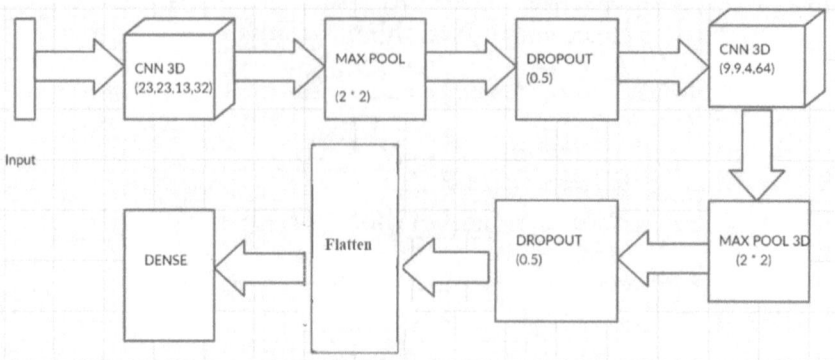

Fig. 10 Block diagram of CNN

to play. A special care has to be taken at the time of converting the multidimensional tensor into a single-dimensional tensor.

The details of the layers used in the 3D CNN model for classification are summarized as follows (Fig. 11). In the first layer, a convolution with filter size 32 has been used with a kernel size of (3, 3, 3). The input image taken is of the shape (25, 25, 15, 1). In the second layer, 3D max-pooling operation has been used, the size of the pooling filter being (2, 2, 2). In the third and fourth layer, batch normalization with 50% of dropout has been used, respectively. The process is repeated again, now taking a convolution of size 64 and kernel size (3, 3, 3) in the fifth layer, followed by 3D max-pooling operation with pooling filter size (2, 2, 2) in the sixth layer. Consequently, in the seventh and eighth phase, batch normalization with 50% of dropout has been used. Following the sequence, the flatten layer is used thereafter, and subsequently, three dense layers have also been used. Among these three dense layers, the first two has a size of 256 units, and the third one is equal to the output unit with softmax as the activation function.

4.2   Significance of 3D CNN in HSI

Hyperspectral images exist in the form of 3D cubes. First of all, the hyperspectral images should be broken in the form of 3D patches, and with the help of 3D kernels, these patches generate different feature maps. Thereafter, these 3D kernel functions collect information from the spatial and spectral bands. In this case, both aspects of spatial and spectral information are extracted from hyperspectral imaging with the help of a satellite sensor. Then the classifier appears to be functioning properly, efficiently, and accurately. As a natural corollary, the 3D CNN functions clearly to classify different classes of a dataset, generating richer feature maps (Fig. 12).

```
Layer (type)                      Output Shape              Param #
=================================================================
conv3d_1 (Conv3D)                 (None, 23, 23, 13, 32)    896

max_pooling3d_1 (MaxPooling3      (None, 11, 11, 6, 32)     0

batch_normalization_1 (Batch      (None, 11, 11, 6, 32)     128

dropout_1 (Dropout)               (None, 11, 11, 6, 32)     0

conv3d_2 (Conv3D)                 (None, 9, 9, 4, 64)       55360

max_pooling3d_2 (MaxPooling3      (None, 4, 4, 2, 64)       0

batch_normalization_2 (Batch      (None, 4, 4, 2, 64)       256

dropout_2 (Dropout)               (None, 4, 4, 2, 64)       0

flatten_1 (Flatten)               (None, 2048)              0

dense_1 (Dense)                   (None, 256)               524544

dense_2 (Dense)                   (None, 256)               65792

dense_3 (Dense)                   (None, 16)                4112
=================================================================
Total params: 651,088
Trainable params: 650,896
Non-trainable params: 192
```

Fig. 11 Model overview

With the application of 3D CNN and kernel size in the hyperspectral cubes, the extraction of better features of an image can be noticed as compared to what 2D CNN does [14]. Hyperspectral imaging basically contains two parameters within it: one is the spectral bands and the other is spatial information. Every spectral band and spatial position in a 3D CNN has its own capabilities to discriminate every feature clearly in a hyperspectral image cube, whereas a 2D CNN alone cannot discriminate the features of spectral dimensions [15].

To implement 3D CNN for the classification of different classes in a hyperspectral image, the captured hyperspectral image has to be normalized first. Basically, the number of bands has to be decreased with the help of linear and nonlinear dimension reduction techniques in this process. After that, the original empty spectrum signature can be obtained from the hyperspectral image. Next, the contiguous range of pixels $P \times P$ has to be sorted. After getting sorted contiguous pixels, the 3D CNN labels the classes at the time of training the model using the stochastic gradient algorithm [16]. The product of spatial dimension and the depth dimension using a 3D kernel enables

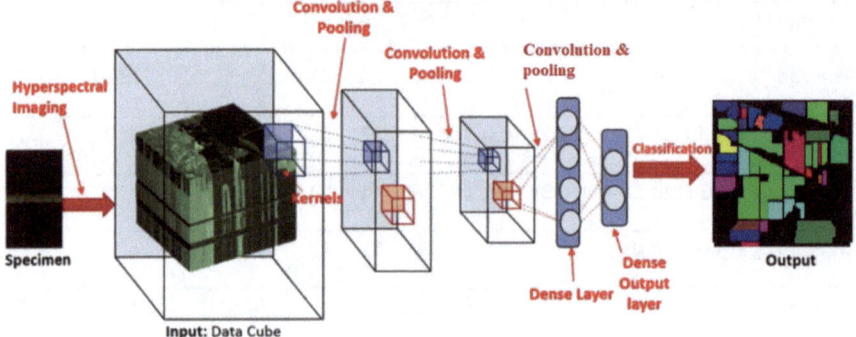

Fig. 12 Convolution neural network

the model to learn the features in a better way. In Eq. 1, Φ denotes the activation function; p, q, r indicate the spatial spotlight of the nth feature map in the mth layer; b denotes the bias function; $w_{m,n,\tau}$ indicates the weight of the 3D kernel; d_{l-1} denotes the number of feature maps in an image of $l-1$th layer; $2\delta-1$ is the width of the kernel, and the kernel height size is given by $2\gamma+1$ [14].

$$a_{m,n}^{p,q,r} = \phi\left(b_{m,n} + \sum_{\tau=1}^{d_{l-1}} \sum_{\lambda=-\eta}^{\eta} \sum_{\rho=-\gamma}^{\gamma} \sum_{\sigma=-\delta}^{\delta} W_{m,n,\tau}^{\sigma,\rho,\lambda} \times a_{m-1,\tau}^{p+\sigma,q+\rho,r+\lambda}\right) \tag{1}$$

4.3 Parameters in CNN

The number of parameters in a given layer refers to the count of learnable elements that are spread in that particular CNN layer. Within a model, there are many layers like the input layer, hidden layers, and output layer, with many neurons present in each layer. In each neuron, we get different weights during the training process because of the changes done through the back-propagation process. Here parameters are nothing but these weights associated with each neuron [1, 17, 18]. The calculation of the learnable parameters o f CNN is presented below.

(i)   **Input Layer**: It only provides the input image shape, and there exist no learnable parameters here.
(ii) **Convolutional Layer**: Each convolutional layer gathers information from the previous layer with some learnable parameters. The learnable parameter can be calculated here by using the formula $((W * H * F) + 1) * N)$, with W as the width, H as the height of this layer, F as the filter size of the previous layer, N as the filter size of the current layer, and 1 as the bias term.

(iii) **Pooling Layer**: In the pooling layer, there is no presence of a back-propagation algorithm in the pooling process. Thus, there is no learnable parameter, or the learnable parameter value is zero.

(iv) **Fully Connected Layer**: Here in this layer, we find more number of learnable parameters than the other layers of the model, the reason being every neuron is connected with each other. The learnable parameter is given by $((P * R) + 1) * P)$, with P as the number of neurons present in the present layer, R as the number of neurons present in the previous layer, and 1 as the bias term.

(v) **Softmax Layer**: The learnable parameters in this the layer can be calculated in the same way as in a fully connected layer, with the same formula and with the same representations [19].

5 Dataset Description

As mentioned earlier, three different hyperspectral datasets have been chosen for carrying out the task of classification, i.e. Indian Pines, Pavia University, and Salinas, the details of which are given below. Figure 13, Fig. 14 and Fig. 15 shows the individual classes with the total number of samples present in each dataset, i.e. Indian Pines, Salinas, and Pavia University, respectively.

(i) **Indian Pines**: Indian Pines dataset consists of 224 spectral bands and 145 × 145 pixels, having one-third components as forests and two-thirds components as agricultural land as its. The ground truth contains 16 classes, some of them including alfalfa, corn, grass pasture, oats, soyabean, wheat, woods, stone steel towers, etc. [18]. The scene contains two main double-lane highways, a railway line, few low density houses, and small roadways. Indian Pines dataset has been obtained from Northern-Western Indiana test site with the help of infrared imaging spectrometer sensor.

(ii) **Pavia University**: Pavia University dataset contains 103 spectral channels in the wavelength of 430–860 nm and 610 × 610 pixels. The ground truth image contains nine classes, some of them including tiles, water, meadows, bare soil, asphalt, etc. Pavia University dataset has been obtained from Northern Italy by using reflective optics system imaging spectrometer (ROSIS) sensor with a geometric resolution of 1.3 m [20]. It can be observed that the rejected samples in the scene are shown as strips of black regions.

(iii) **Salinas Valley**: Salinas dataset contains an image of 512 × 217 spatial dimensions and 224 spectral bands in the wavelength ranging from 360 to 2500 nm, the image being accessible as a sensor radiation data. It consists of 16 classes like broccoli green weeds, stubble, celery, vinyard, corn sensed green weeds, grapes, etc. Salinas dataset has been obtained from Salinas Valley, California, with the help of AVIRIS sensor. The total number of samples in each of the datasets is Indian Pines (10,249), Salinas (54,129), and Pavia University (7446).

Fig. 13 Indian Pines

SL	Classes	Samples
1	Alfalfa	46
2	Corn-notill	1428
3	Corn-mintill	830
4	Corn	237
5	Grass-pasture	483
6	Grass-trees	730
7	Grass-pasture-mowed	28
8	Hay-windrowed	478
9	Oats	20
10	Soyabean-notill	972
11	Soyabean-mintill	2455
12	Soyabean –clean	593
13	Wheat	205
14	Woods	1265
15	Building-Grass-Trees-Drives	386
16	Stone-Steel-Towers	93

6 Result Analysis

After successfully building the CNN model, the model was made to run for finding the accuracy of the model in predicting the different samples of the respective datasets. For the purpose, the model was run with 100 epochs and with a batch size of 128. The accuracy achieved by the model after the successful running of the epochs was 99.81%, with the validation accuracy of 99% for each of the datasets. Accuracy and Loss graph obtained from three different Hypserspectral image datasets is shown in Fig. 16, Fig. 17 and Fig. 18.

Fig. 14 Salinas Valley

SL	Classes	Samples
1	Brocoli_green_weeds_1	2009
2	Brocoli_green_weeds_2	3726
3	Fallow	1976
4	Falow_rough_plow	1394
5	Fallow_smooth	2678
6	Stubble	3959
7	Celery	3579
8	Grapes_untrained	11271
9	Soil_vinyard_develop	6203
10	Corn_seesced_green_weeds	3278
11	Lettuce_romaine_4wk	1068
12	Lettuce_romaine_5wk	1927
13	Lettuce_romaine_6wk	916
14	Lettuce_romaine_7wk	1070
15	Vinyard_untrained	7268
16	Vinyard_vertical_trellis	1807

6.1 Ground Truth Image Significance

The function of the ground truth image is to capture the information of a particular location. When information for a hyperspectral image is gathered from the satellite images, the information undergoes some distortion because of the absorption of poor atmospheric conditions and obstructions [21]. To overcome the problems originating from distorted information, the ground truth image information is taken as a reference for help. The accuracy of the ground truth image remains at its level best. But in the effort of measuring the level of accuracy of our model using the testing dataset, the precision accuracy of the model somehow appears to be less. Ground truth data is generally obtained by visiting the respective site or location, surveying the particular location, and measuring different properties and features of the site, for example, how much of the area is expanded by the forests, the span of the

SL	Classes	Samples
1	Water	824
2	Trees	810
3	Asphalt	816
4	Self-Blocking-Bricks	808
5	Bitumen	808
6	Tiles	1260
7	Shadows	476
8	Meadows	824
9	Bare-Soil	820

Fig. 15 Pavia University

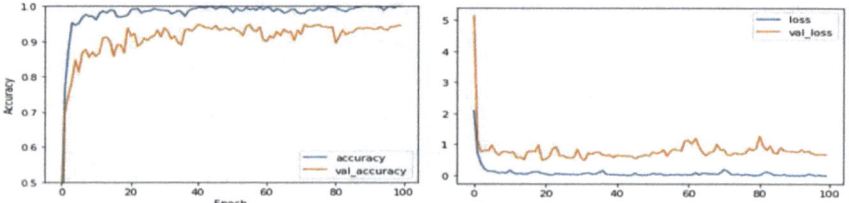

Fig. 16 Indian Pines: accuracy graph (left), loss graph (right)

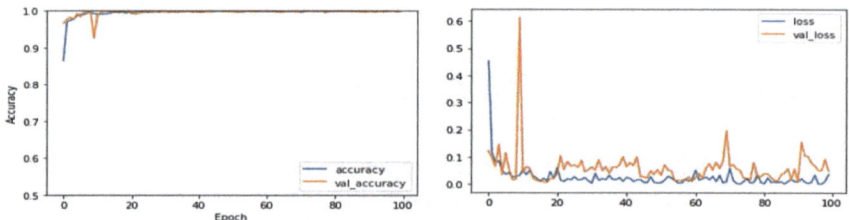

Fig. 17 Pavia University: accuracy graph (left), loss graph (right)

water area, vegetation, other classes of land, the spread of buildings, etc. To classify different classes of the satellite image, the ground truth image is taken, where each pixel of the satellite image is compared with the corresponding pixels of the ground truth image. Manual verification of the spot is very much difficult and laborious since it requires strenuous efforts to do the job, and very often it may hinder the

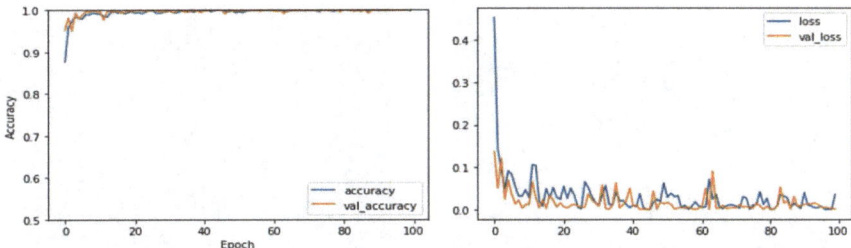

Fig. 18 Salinas: accuracy graph (left), loss graph (right)

progress towards perfection. Finding some geographic location sometimes becomes very much challenging to obtain the ground truth image record of the earth's surface. To counter these challenging hindrances, deep learning techniques are adopted for the classification of different classes [11].

False Colour Image: For a false colour image, lets us visualize a complete and composite picture of the wavelength of a particular range of a band. It shows the multispectral image which consists of different wavelengths other than red, green, and blue bands, visible to the human eye. False colour image uses at least one non-visible wavelength band, and after the non-visible wavelength is received, it focuses on visible wavelength bands like red, green, or blue. For this reason, the image is not seen in its proper colour at times. For example, the green grass may not be appearing green in colour all the time. The False colour image of Pavia University dataset is shown in Fig. 19.

The ground truth image and the predicted image for the three datasets after successful running of the model are shown in Fig. 20, Fig. 21 and Fig. 22.

Fig. 19 False colour image of Pavia University [22]

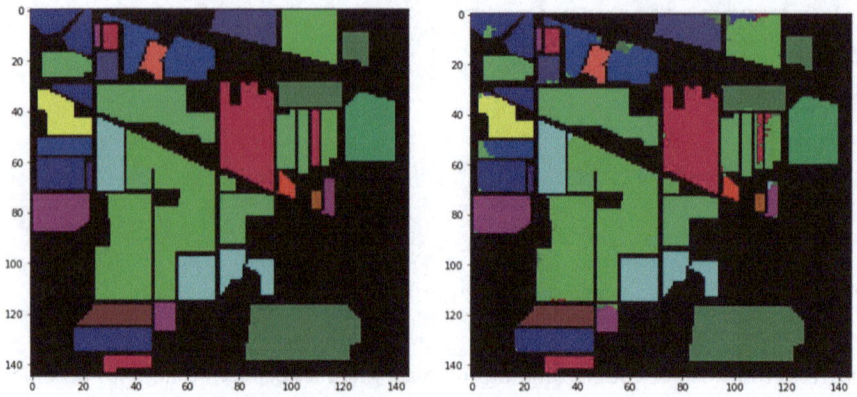

Fig. 20 Indian Pines: ground truth image (left), predicted image (right)

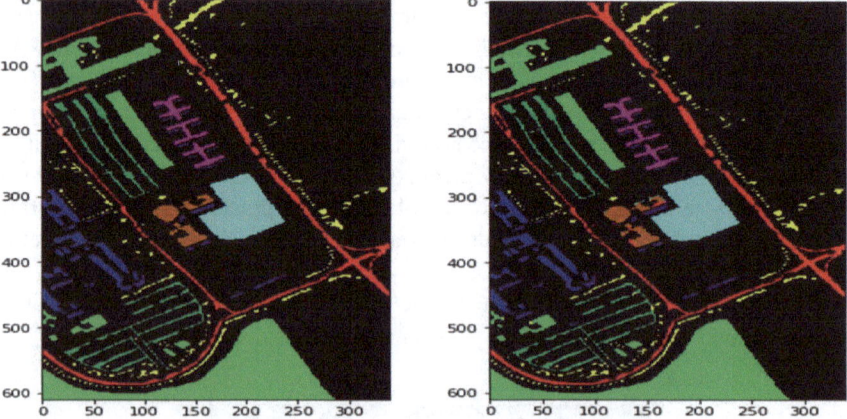

Fig. 21 Pavia University: ground truth image (left), predicted image (right)

6.2 Class Wise Accuracy

The accuracy of a model signifies how correctly or precisely a model is able to classify or predict a particular dataset or class, as in this case. When the classes of a particular dataset are roughly balanced in equal sizes, then the accuracy metric is used. Precision is used when the dataset is totally imbalanced in order to predict the accuracy level of the model. Recall indicates the true positive rate of the model. The $F1$-score tries to make synchronization between precision and recall. Meanwhile, the equations for calculating the above-mentioned parameters are given as follows.

$$\text{Accuracy} = \frac{\text{TP} + \text{TN}}{\text{TP} + \text{FP} + \text{TN} + \text{FN}} \tag{2}$$

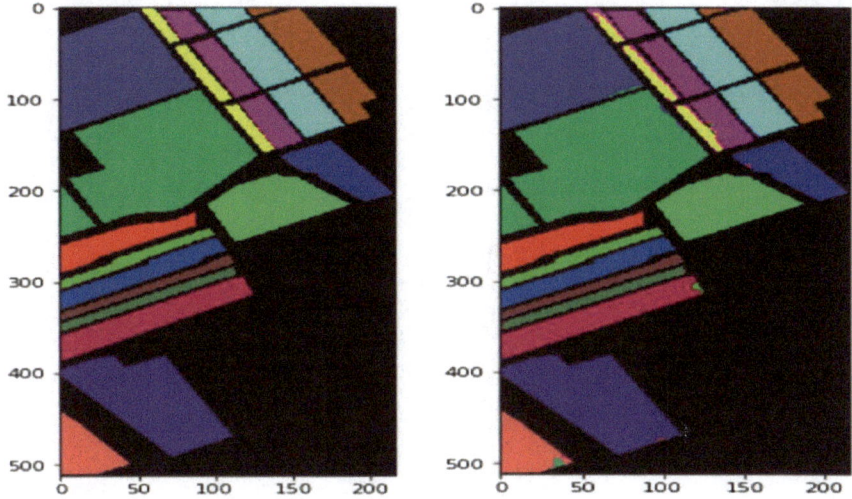

Fig. 22 Salinas: ground truth image (left), predicted image (right)

$$\text{Precision} = \frac{\text{TP}}{\text{TP} + \text{FP}} \tag{3}$$

$$\text{Recall} = \frac{\text{TP}}{\text{TP} + \text{FN}} \tag{4}$$

$$F_1 = 2 \times \frac{\text{precision} \times \text{recall}}{\text{precision} + \text{recall}} = \frac{2 \times \text{TP}}{2 \times \text{TP} + \text{FP} + \text{FN}} \tag{5}$$

6.3 Confusion Matrix

With the help of the confusion matrix, the performance of the classification model can be measured on test data. The confusion matrix has four basic terms, namely true positive, true negative, false positive, and false negative. True positive identifies the real existence of the entity as it is to be verified and identified. To be more particular, the model here tries to identify the presence of anything which is truly present which is to be proved correct. True negative identifies the non-existence of the entity as is to be proved and indicated. To be more particular, the model tries to identify the absence of anything which is not present that is to be proved correct. False positive shows the existence of an entity predicted by the model which is not real. False negative refers to the non-existence of the entity. To be more precise, it may be said that the prediction of the non-existence of the entity is wrong (Fig. 23).

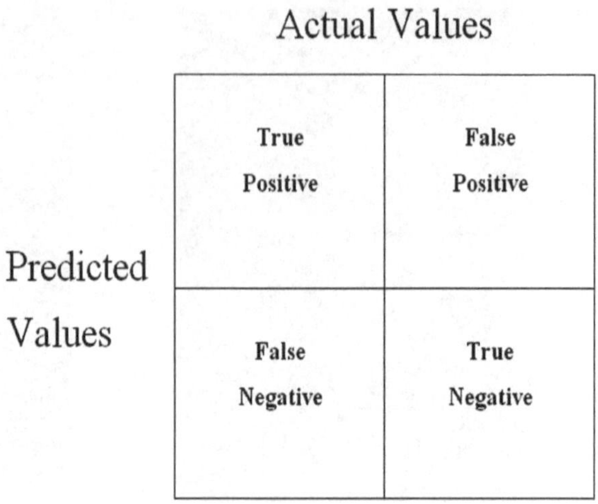

Fig. 23 Confusion matrix

Actual Values

	True Positive	False Positive
Predicted Values	False Negative	True Negative

The equations for calculating the kappa accuracy (KA), overall accuracy (OA), and Average accuracy (AA) are given as follows.

$$KA = \frac{accuracy_{total} - accuracy_{random}}{1 - accuracy_{random}} \tag{6}$$

$$OA = \frac{\left(K \times accuracy_{average} + (2 - K)\right)}{2} \tag{7}$$

$$AA = \frac{2 \times OA - 1}{K + 1} \tag{8}$$

where K is the target class. The accuracy values obtained for the three datasets after classification are shown in Table 1. The highest accuracy values for each of the dataset are shown in bold. The table shows that Indian Pines obtain the highest accuracy values compared to the other two datasets. The confusion matrix for the three datasets so obtained is shown in Fig. 24, Fig 25 and Fig 26.

Table 1 Accuracy obtained using the datasets used

Dataset	Indian Pines	Pavia University	Salinas Valley
Training size (%)	10	5	5
Kappa accuracy	**99.80**	98.90	98.91
Overall accuracy	**98.20**	98.15	98.19
Average accuracy	**99.86**	98.26	98.45

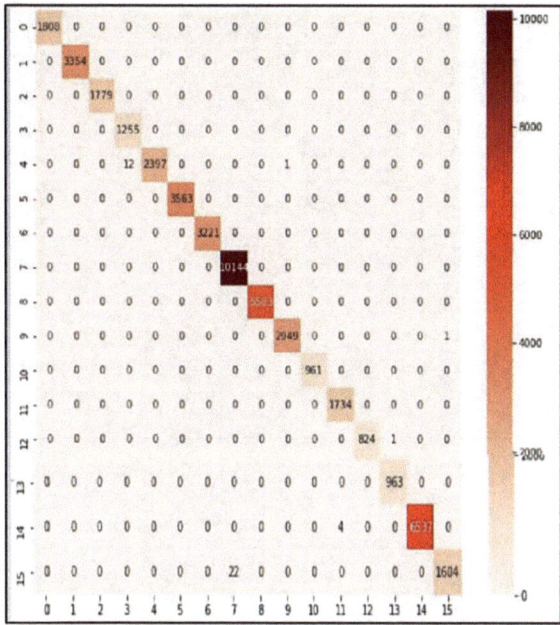

Fig. 24 Indian Pines

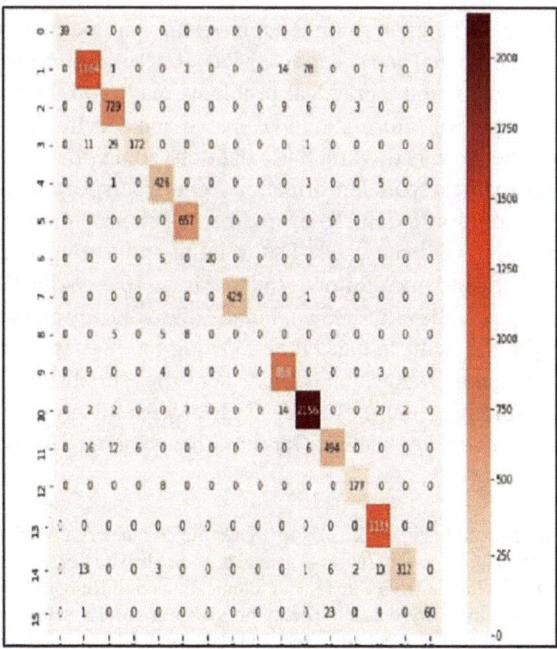

Fig. 25 Salinas

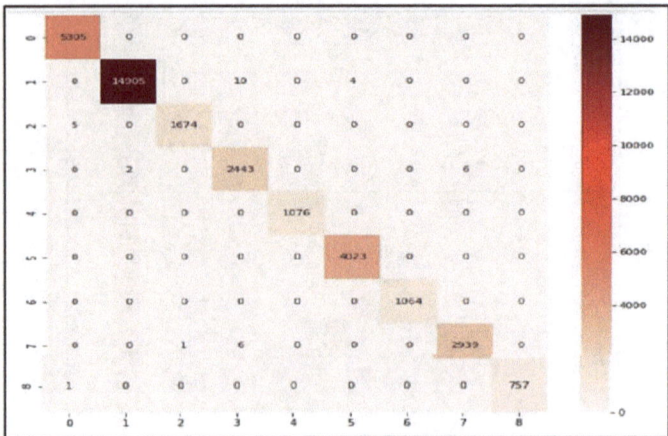

Fig. 26 Pavia University

7 Conclusion

Hyperspectral imaging has been evolved as a hopeful and adaptable signal processing method in the field of remotely sensed image processing for a broad variety of application domains. In this process, dimensional reduction followed by classification of hyperspectral images helps in finding latent features of the materials present on earth's surface, both in spatial and spectral aspects. If 2D convolution neural network is used on these datasets, some practical problems may crop up. In hyperspectral image, there are two sets of information like spatial and spectral information. If 2D convolution neural network is used on these datasets, some practical problems may crop up, and it only extracts spatial information from hyperspectral image but it does not cover the spectral information. To overcome this shortcoming, 3D convolution neural network (CNN) has been used. The training and testing sizes used for the three datasets, respectively, are as follows: Indian Pines (10%, 90%), Pavia University (5%, 95%), and Salinas Valley (5%, 95%). Finally, the accuracy and validation levels reached to about 99.5%, with Indian Pines showing the highest accuracies among the datasets. These attempts show that HSI can come up with distinct methodologies for concurrently estimating the presence of various useful properties that can be used in numerous potential applications for the enhancement of evolving computing technology.

Acknowledgements The authors would like to thank the Pursue's University MultiSpec site through which the Indian Pines dataset was available, Kaggle site for providing the Salinas Valley dataset, and Prof. Gamba from the Telecommunications and Remote Sensing Laboratory, for providing the Pavia University dataset. The authors also gratefully acknowledge the helpful comments and suggestions of the associate editors and reviewers, which have improved the quality of the presentation.

References

1. S. Li, W. Song, L. Fang, Y. Chen, P. Ghamisi, J.A. Benediktsson, Deep learning for hyperspectral image classification: an overview. IEEE Trans. Geosci. Remote Sens. (2019)
2. M.J. Khan, H.S. Khan, A. Yousaf, K. Khurshid, A. Abbas, Modern trends in hyperspectral image analysis: a review. IEEE Access (2018)
3. https://www.examrace.com/Current-Affairs/NEWS-ISRO-Developing-Hyperspectral-Imaging-Earth-Observation-Satellite.htm
4. X. Kang, S. Li, L. Fang, M. Li, J.A. Benediktsson, Extended random walker-based classification of hyperspectral images. IEEE Trans. Geosci. Remote Sens. **53**(1) (2015)
5. Y. Xu, Z. Wu, Z. Wei, Spectral–spatial classification of hyperspectral image based on low-rank decomposition. IEEE J. Sel. Top. Appl. Earth Obs. Remote Sens. (2015)
6. https://www.cis.rit.edu/research/thesis/bs/1999/newland/thesis.html
7. M.E. Paoletti, J.M. Haut, J. Plaza, A. Plaza, Deep learning classifiers for hyperspectral imaging: a review. ISPRS J. Photogramm. Remote Sens. (2019)
8. K. Ose, T. Corpetti, L. Demaigistri, Optical remote sensing of land surface (2016)
9. M. Mateen, J. Wen, Nasrullah, M.A. Akbar, The role of hyperspectral imaging: a literature. Int. J. Adv. Comput. Sci. Appl. (IJACSA) (2018)
10. J. Han, J. Pe, A volume in the Morgan Kaufmann series in data management systems (2012)
11. https://medium.com/@iamvarman/how-to-calculate-the-number-of-parameters-in-the-cnn-5bd55364d7ca
12. C. Wang, W. Fu, H. Huang, J. Chen, Isomap-based three-dimensional operational modal analysis (2020)
13. S.H. Shabbeer Basha, S.R. Dubey, V. Pulabaigari, S. Mukherjee, Impact of fully connected layers on performance of convolutional neural networks for image classification. Neurocomputing (2019)
14. S.K. Roy, G. Krishna, S.R. Dubey, B.B. Chaudhuri, Hybrid SN: exploring 3D-2D CNN feature hierarchy for hyperspectral image classification (2019)
15. C. Yu, R. Han, M. Song, C. Liu, C.-I. Chang, A simplified 2D-3D CNN architecture for hyperspectral image classification based on spatial–spectral fusion. IEEE J. Sel. Top. Appl. Earth Obs. Remote Sens. (2020)
16. L. Chen, Z. Wei, Y. Xu, A lightweight spectral-spatial extraction and fusion network for hyperspectral image classification (2020)
17. https://mc.ai/understanding-and-calculating-the-number-of-parameters-in-convolution-neural-networks-cnns/
18. https://rslab.ut.ac.ir/data
19. https://towardsdatascience.com/understanding-and-calculating-the-number-of-parameters-in-convolution-neural-networks-cnns
20. https://www.kaggle.com/abhijeetgo/paviauniversity
21. https://towardsdatascience.com/an-image-processing-tool-to-generate-ground-truth-data-from-satellite-images-using-deep-learning-f9fd21625f6c
22. Z. Li, X. Tang, W. Li, C. Wang, C. Liu, J. He, A two-stage deep domain adaptation method for hyperspectral image classification (2020)

Detection of Coronavirus (COVID-19) Using Deep Convolutional Neural Networks with Transfer Learning Using Chest X-Ray Images

Sayantan Ghosh and Mainak Bandyopadhyay

Abstract The novel coronavirus, COVID-19, has caused the greatest crisis of human civilization. Originated in Wuhan, China, it has gradually spread all over the world. The deadly virus has been threatening human mankind for the last three months. In the surge of COVID-19 pandemic, precise and immediate detection of COVID19 cases will immensely help in the testing for COVID-19 detection. Therefore, there is a substantial need for an auxiliary automated detection system which will help in immediate testing. Since the virus is targeting the human respiratory system initially, chest X-ray images will come out to be a useful feature for COVID-19 detection in an early stage. With the help of deep convolutional neural networks (CNN), it is possible to get promising results. In this paper, a new fine-tuned deep CNN model has been proposed to generate precise and accurate diagnostics for binary classification (COVID-19 positive vs COVID-19 negative) using raw chest X-ray images. The detailed model architecture and accuracy metrics are presented which are obtained from fivefold cross-validation trained over 1300 image samples. The proposed model gained a validation accuracy of 99.39394% which is superior as compared to other studies in this domain.

Keywords COVID-19 · VGG16 · Lungs X-ray radiograph · Transfer learning

1 Introduction

The COVID-19 (coronavirus disease 2109) infection, which began generating headlines, originated in Wuhan, China, in December 2019, has expanded rapidly all over the world and became a pandemic [1]. This is known as COVID-19, and the causing virus is named as SARS-CoV-2 by the International Committee of the

S. Ghosh (✉) · M. Bandyopadhyay
School of Computer Science and Engineering, Kalinga Institute of Industrial Technology
Deemed-to-be-University, Bhubaneswar, Odisha, India

M. Bandyopadhyay
e-mail: mainak.bandyopadhyayfcs@kiit.ac.in

© The Author(s), under exclusive license to Springer Nature Singapore Pte Ltd. 2021
M. Bandyopadhyay et al. (eds.), *Machine Learning Approaches for Urban Computing*,
Studies in Computational Intelligence 968,
https://doi.org/10.1007/978-981-16-0935-0_4

63

Taxonomy of Viruses (ICTV). It belongs to a virus family causing several diseases ranging from "severe acute respiratory syndrome(SARS-CoV), middle east respiratory syndrome(MERS-CoV) causing deaths and acute respiratory syndrome in humans" [2]. The new species of coronavirus which took a surge in March 2020 has the capability of a person-to-person transmission via respiratory droplets which is the reason for the rapid spreading [3]. It has been presumed that the virus mainly affects animals first, especially snakes and bats, and then humans due to its zoonotic nature [3, 4].

It has been observed that in the majority, 98% of cases are mild conditions, whereas only 2% of cases are serious or critical. As of the current situation, more than five million people are infected, and there are over 350,000 deaths all over the world [5]. COVID-19 has been declared as a Public Health Emergency of International Concern (PHEIC) by WHO on January 30 [6]. Researchers as well as medical healthcare professionals are finding new things about this virus every day. Therefore, some common symptoms with seasonal flu have been noticed in the patients, which include shortness of breath, cough, fever, sore throat and fatigue. Sometimes, these symptoms become more severe in some patients like multi-organ failure, septic shock, severe chest pain and death [7, 8]. The most advanced test mechanism for SARS-CoV2 detection is real-time reverse transcription polymerase chain reaction (RT-PCR) [9], which is highly specific and the sensitivity reported as low as 60–70% and high as 95–97% [10]. Also due to its large cost and complexity, as it is an RNA-extracting machine, it needs highly trained professionals and advanced laboratory equipment [11]. Lack of laboratory facilities causes a substantial amount of delay for the precise diagnosis of suspected patients, which is a crucial problem as the pandemic is hard to control and for its rapidly expanding nature. Radiological raw images of the chest such as X-ray and computed tomography(CT) of the lungs play a lenient role in faster and early detection of COVID-19 disease. As of current findings, CT images stand to be an effective methodology for detecting COVID-19 pneumonia, which also can be used for RT-PCR. Moreover, a study depicts that 30% of the positive cases never showed recognizable symptoms and changes in CT images. 20% of the reported cases showed symptoms in the hospital, which suggests that a major percentage of COVID-19 carriers seemed to be asymptomatic [12]. Furthermore, it has been found that a recovered patient can also show symptoms and found positive test results urging a need for more accurate imaging analysis. "Based on the research of Zu et al. [13] and Chung et al. [14], it is seen that 33% of COVID-19 chest CTs have a tendency of rounded lung opacities". In Fig. 1, chest X-ray images were taken at days 1, 4, 8 and 13 for a 73-aged COVID-19 male patient with aortic insufficiency, and the detailed findings are mentioned [15].

Machine learning and deep learning applications in medical images are one of the most advanced research fields in the health sector as it also acts as an adjacent tool for health workers [16]. Deep learning is also an emerging field of research enabling to create advanced deep learning models to achieve precise and accurate results without the need for feature extraction [17] manually. Applications of deep learning include detection of pneumonia from lung X-ray [18], brain tumour segmentation [19] and lung cancer detection using 3D convolutional networks [20]. Due to a limited

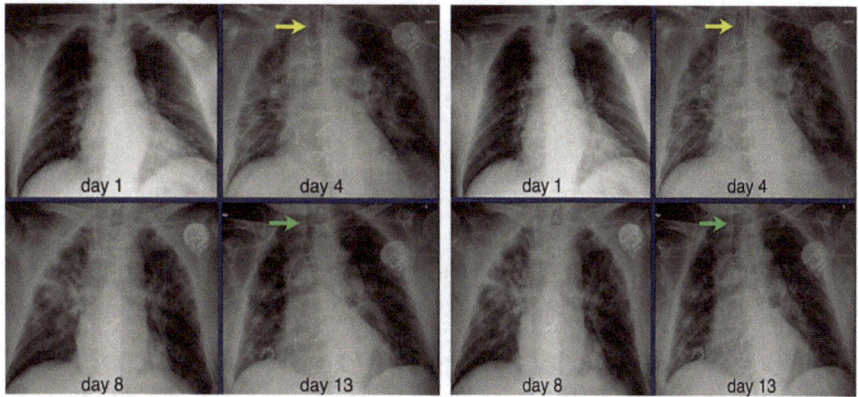

Fig. 1 Day 1 normal findings, day 4 bilateral consolidations intubated, day 8 bilateral consolidation, day 13 extubation

number of radiologists, it is really a challenging task to test and examine a substantial number of X-ray images due to the rapid increase of coronavirus infection. As for the scenario, AI-enabled and deep learning-based automated solutions provide timely assistance to the radiologists, doctors and also helpful to obtain better results [21]. It has the performance on the ImageNet image classification task which was beyond human-level performance which had ten lakhs images in the training phase in 2015, obtaining promising results in the cancer detection of lungs in 2019 [22, 23].

Recently, many deep neural networks have been proposed for the detection of early stages of COVID-19. Ceren Kaya et al. [24] obtained an accuracy of 98% for analysing the COVID-19 from lungs X-ray images using the ResNet-50 model. Wang et al. [25] proposed an architecture of CNN named COVID-Net for the early detection of COVID-19, obtaining 92.4% accuracy in classifying COVID-19 classes, normal and non-COVID pneumonia. Sethy and Behra [26] obtained features and classified from different CNN along with support vector machine (SVM) classifier for the best performance. Tulin Ozturk et al. [27] proposed DarkCovidNet obtaining 98.08% in the binary classification task, whereas 87.02% for the multiclass classification task.

This study aims to build and develop a fine-tuned CNN architecture coupled with transfer learning, which will assist in the early detection of COVID-19 pandemic and to develop deep learning lungs X-ray images. The overall architecture requires raw radiological images of chest X-ray, and it provides a probability of COVID-19 positive of the X-ray image. We have trained our model with 141 COVID-19 positive chest X-ray images, obtained briskly. We have used VGG16 as our base model with fine-tuning and transfer Learning.

2   Materials and Methods

2.1   Data Set Collection and Findings

Chest X-ray image samples are collected available in the GitHub public repository, which was developed by Cohen JP [28]. The database is regularly updated by a group of researchers. Currently, there are a total of 141 samples of COVID-19 positive. The data set contains 57% of male, 32% female, and 12% are other cases. In the data set, metadata is given, and the age distribution is provided. The normal X-ray images of the lung are collected from the public data set repository of Kaggle [29]. There are a total of 1341 images in the X-ray data set, which are all resized to 224 × 224 pixels. In Fig. 2, normal patient X-ray images are provided, and in Fig. 3 COVID-19 positive X-ray samples are provided.

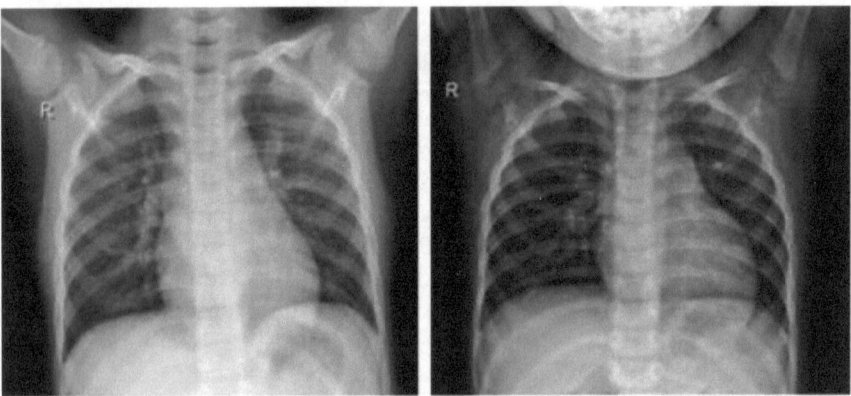

Fig. 2 X-ray samples of normal patient

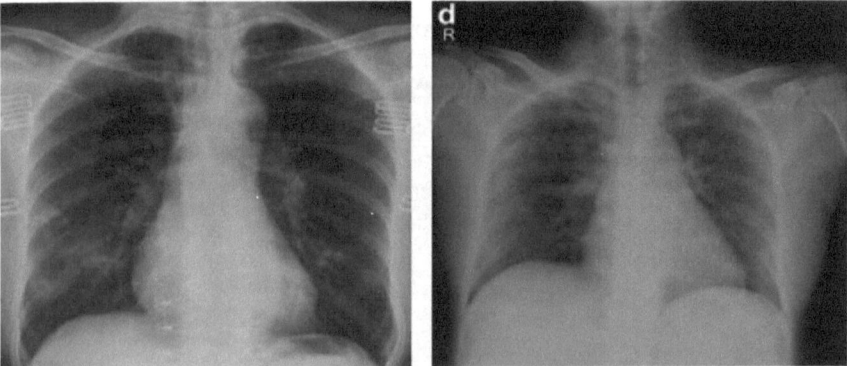

Fig. 3 X-ray samples of COVID-19 positive Image

2.2 VGG16 Architecture

VGG16 is one of the best dense convolutional network models proposed by K. Simonyan and A. Zisserman mentioned in the paper titled "Very Deep Convolutional Networks for Large-Scale Image Recognition" [30]. It is one of the most accurate models obtaining 92.7% accuracy in ImageNet large-scale visual recognition challenge (ILSVRC) consisting of more than 14 million images belonging to 1000 classes. VGG16 is an improved version of AlexNet [31] by replacing the kernel filters of 3 × 3. It has 13 convolution layers with three dense layers. VGG16 is one of the dense networks consisting of 138 million parameters. The model takes an input image (RGB) of a fixed size of 224 × 224 pixels. It uses 1 × 1 convolution filters for linear transformation of the input filters. Spatial pooling is performed by five max-pooling layers with a window size of 2 × 2 and a stride number of 2.

VGG16 consists of three fully connected convolution layers which have varying depths depending upon the architecture. The first two consist of 4096 channels each, and the third contains 1000 channels. Softmax activation function is used in the final layer. The mathematical formulation of 2D convolution is given in Eq. (1).

$$y[i, j] = \sum_{m=-\infty}^{\infty} \sum_{n=-\infty}^{\infty} h[m, n] \cdot x[i - m, j - n] \tag{1}$$

Here, x represents the input image matrix which is to be convoluted with the (3 × 3) kernel matrix h to result in a new feature map. Here y represents the output image, and the indices i and j are related with the image matrices, while m and n deal with the kernel. The indices m and n range from -1 to 1. We have used a stride matrix by (1,1).

2.3 Proposed Architecture of VGG16 with Transfer Learning

Deep learning is one of the sub-fields of machine learning, and the growth of deep learning has revolutionized in the domain of artificial intelligence(AI). Deep learning models have been used in extracting features from images to draw meaningful insights. The CNN architecture has been named after the mathematical convolution operation. The convolution layers are used to extract meaningful features using the input filters to create the feature map. Stacking these convolution layers, a typical CNN model is formed, which can be used for image classification and segmentation in medical data.

Pre-trained deep CNN models are available, so developing a deep network model from root, a most robust approach, is to use a proven pre-trained model. In the study of medical imaging, data sets availability is one of the most crucial problems that a data scientist often faces. Normally, to train a CNN model and to extract enough information from the data, a dense model requires a large amount of data. In this case, transfer learning [32] and fine-tuning of the pre-trained model come into play because it allows the training of deep CNN models with a limited number of data resources. Transfer learning is the methodology to reuse and tune a pre-trained model. Nevertheless, it is one of the growing research interests in the field of deep learning [33].

So, in this study, we have used a deep CNN based on VGG-16 for the detection of COVID-19 using two classes belonging to normal patients and COVID-19 positive patients. In addition to this, we have implemented the transfer learning technique that has been utilized by using ImageNet data for overcoming the insufficiency of data. The schematic representation of the VGG-16 CNN model has been depicted in Fig. 4. We have frozen the top layers of the VGG-16 model fine-tuning the fully connected layers using transfer learning. We have used average pooling. The proposed and fine-tuned final block architecture is provided in Fig. 4.

For the pooling layer, the max-pooling method has been used. Max-pooling is used to reduce the input shape dimension and allow assumptions to be made about features. All the convolution, max-pooling layers and a number of trainable parameters of the proposed VGG16 are shown in the batch size and learning rate which are 8 and $1e^{-3}$, respectively. After freezing the top layer for the training phase, it consists of 14 million parameters (Table 1)

3   Experimental Set-up and Results

Here Python 3.6.5 and Kaggle eCloud GPU (P100) are used for the training purpose of our VGG16 deep neural model. A binary classification model which can accurately classify X-ray images into two classes, COVID-19 positive and normal cases, is developed. The overall performance is computed using a fivefold cross-validation strategy. For each fold, the total feature space is split into 80% of data that has been used for training purpose and 20% for the validation purpose. The cross-validation strategy has been shown in Fig. 5. In each fold, our proposed model is trained for 100 epochs and a total of 500 epochs. The training accuracy, validation accuracy and training validation losses for fivefolds of the VGG16 model are shown in Figs. 6 and 7.

The details of the confusion matrices (CM) of each fold for two classes (COVID-19 and normal) have been displayed in Fig. 8.

The details of the performance metrics along with precision score, sensitivity score, specificity score, recall score and F1-score are shown in Table 2.

Comparative Analysis between Pre-trained VGG16 and Fine-tuned VGG16 + Transfer Learning:

The CNN base model has been built using a pre-trained VGG16 model. Further fine-tuning and transfer learning are applied to increase the accuracy. The accuracy and the loss comparison of VGG16 and VGG16 + transfer learning is presented in Fig. 9.

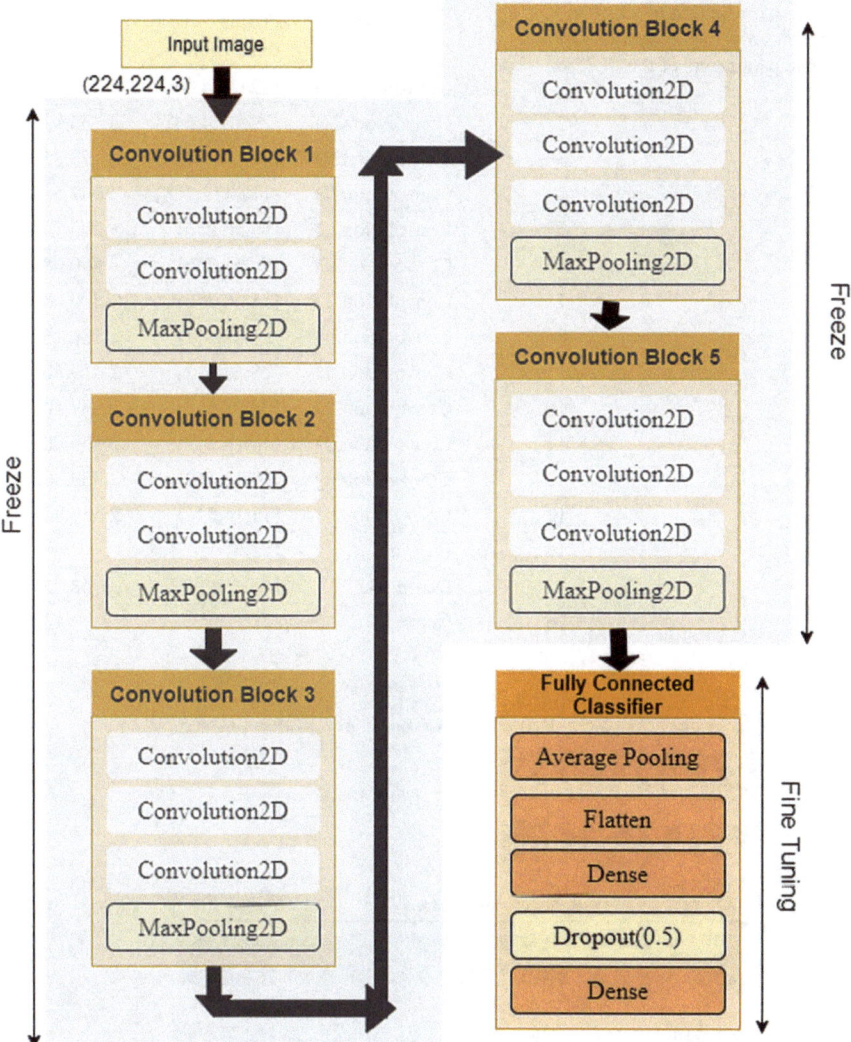

Fig. 4 Fine-tuned VGG16 architecture

The ROC accuracy is coming to be 98%. The ROC curve distribution has shown in Fig. 10. From Table 2, the proposed VGG16 model with transfer learning has achieved an average accuracy of 99.458% for the classification COVID-19 diseases. The average specificity, sensitivity, precision, F1-score and recall score are found to be 99.698%, 97.138%, 97.138%, 97.136% and 97.214%, respectively.

Table 1 Convolution blocks, dimensions and the number of trainable parameters of the model

Layer no	Layer type	Output dimension	Trainable parameters
1	Convolution2D	[224,224,64]	1792
2	Convolution2D	[224,224,64]	36,928
3	Convolution2D	[112,112,128]	73,856
4	Convolution2D	[112,112,128]	147,584
5	Convolution2D	[56,56,256]	295,168
6	Convolution2D	[56,56,256]	590,080
7	Convolution2D	[28,28,256]	590,080
8	Convolution2D	[28,28,512]	1,180,160
9	Convolution2D	[28,28,512]	2,359,808
10	Convolution2D	[28,28,512]	2,359,808
11	Convolution2D	[14,14,512]	2,359,808
12	Convolution2D	[14,14,512]	2,359,808
13	Convolution2D	[14,14,512]	2,359,808
14	Flatten	[512]	0
15	Dense[64]	[64]	32,832
16	Dense [2]	[2]	130

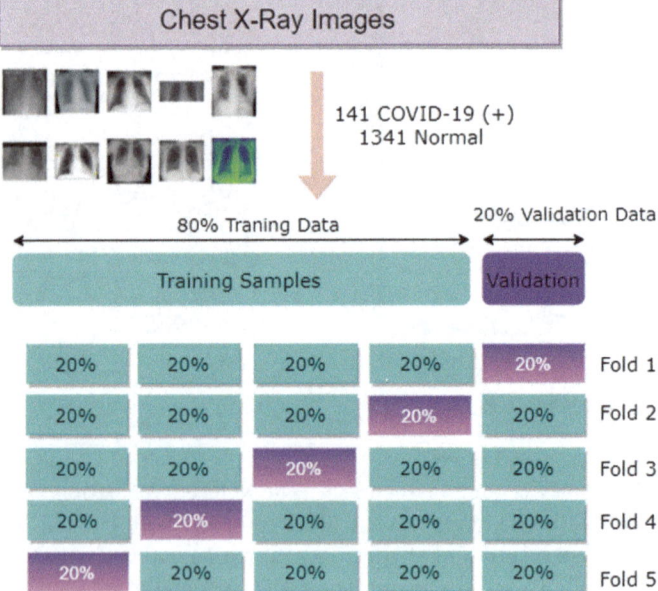

Fig. 5 Representation of the K-fold strategy

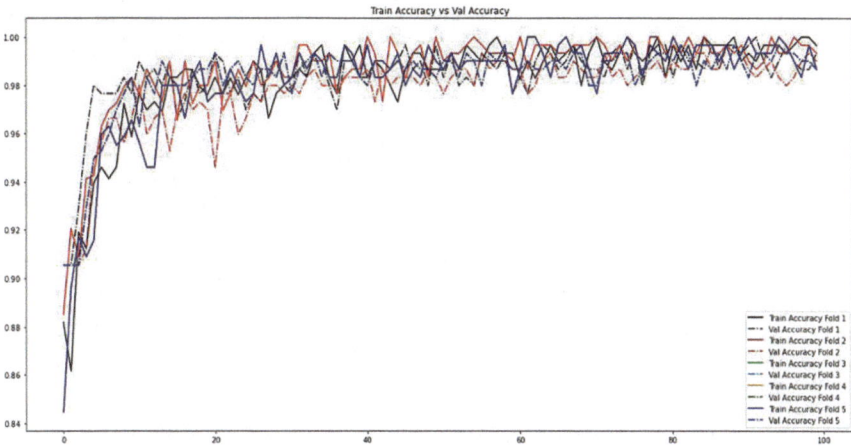

Fig. 6 Training and validation accuracy curve over 100 epochs

4 Discussion

Radiological images of chest X-ray images are being constantly updated by Dr. Cohen for the researchers to make an efficient and accurate model for the early-stage detection of COVID-19. Deep learning models, especially deep CNN, are commonly used for extracting the features which are relevant from the samples of the X-ray images. The normal chest X-ray images are available in public repositories and Kaggle resources. Wang and Wong [25] proposed the architecture of COVID-Net, for the early-stage detection of COVID-19. COVID-Net obtained an accuracy of 92.4%. They have used a sample size of sixteen thousand which are gathered from different public repositories. Ioannis et al. [34] applied transfer learning in the same **COVID**-Net model, and he used a sample of 224 positive X-ray images, 504 normal radiology images and 700 pneumonia. They obtained a 98.75% performance for the two-class classification problem. Another study of Zheng et al. [35] showed that they have achieved 90.8% accuracy by using 313 positive samples of COVID-19 and 229 normal samples. Also, Tulin Ozturk et al. proposed their DarkCovidNet model obtaining 98.08% in the binary classification task. They have used 125 COVID-19 positive image samples and 500 no findings for the purpose [27].

In our proposed study, we have used the base model as VGG16 and applied transfer learning and fine-tuned. We have used 141 positive COVID-19 samples and 1341 normal X-ray images from Kaggle. After fivefold of cross-validation, we have obtained an average accuracy of 99.458% for binary classes, which is more superior in comparison with other studies provided in Table 3.

The proposed model can provide advanced assistance to medical health workers and in the detection of coronavirus. It will also help to reduce the time complexity of testing and the limitation of resources. X-ray images are easily available, and it has crucial information about the patient. However, the only limitation is the small

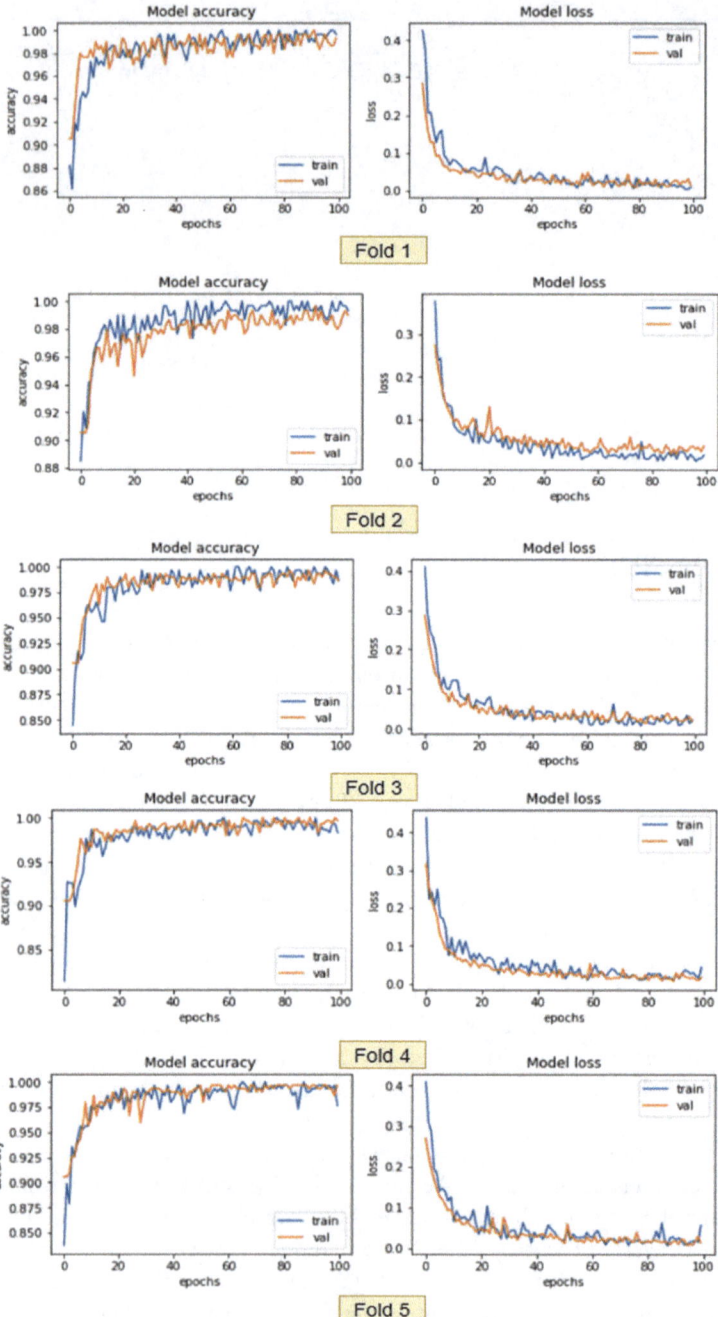

Fig. 7 Representative graph of training and validation accuracy of fivefold

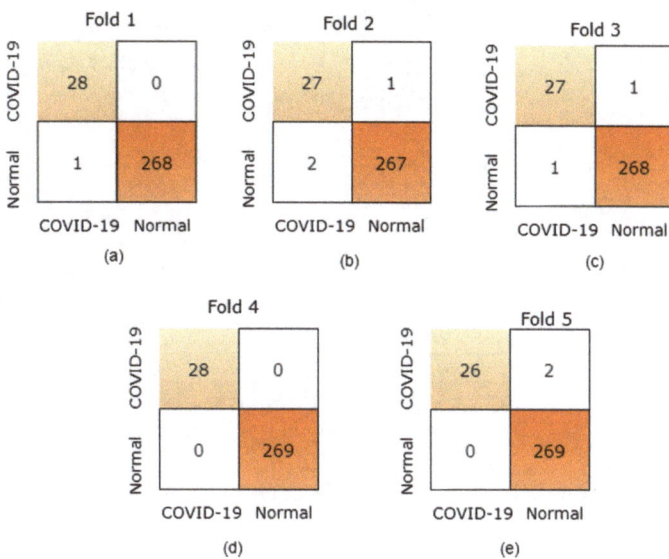

Fig. 8　a Fold 1 CM **b** Fold 2 CM **c** Fold 3 CM **d** Fold 4 CM **e** Fold 5 CM

Table 2 Specificity, sensitivity, precision, F1-score, recall and accuracy values of two classes COVID-19 and normal using VGG16 and transfer learning

5 Folds	Performance metrics(%)					
	Specificity	Sensitivity score	Precision score	F1-score	Recall score	Accuracy
Fold1	99.62	100	100	98.24	96.55	99.66
Fold2	99.25	96.42	96.42	94.73	93.10	98.98
Fold3	99.62	96.42	96.42	96.42	96.42	99.33
Fold4	100	100	100	100	100	100
Fold5	100	92.85	92.85	96.29	100	99.32
Average	99.698	97.138	97.138	97.136	97.214	99.458

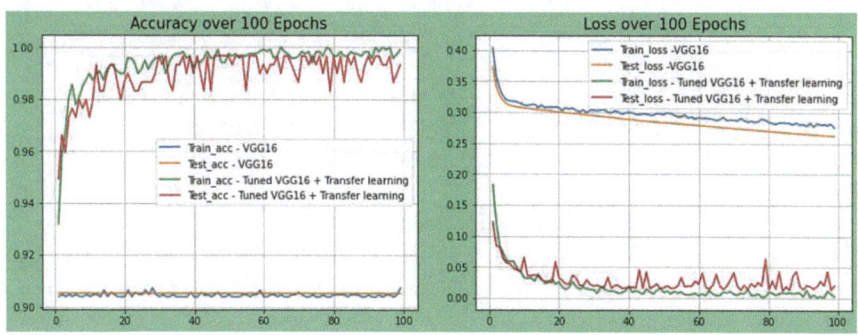

Fig. 9 Training and test accuracy and loss of VGG16 and VGG16 + transfer learning

Fig. 10 ROC accuracy of
VGG16 + transfer learning

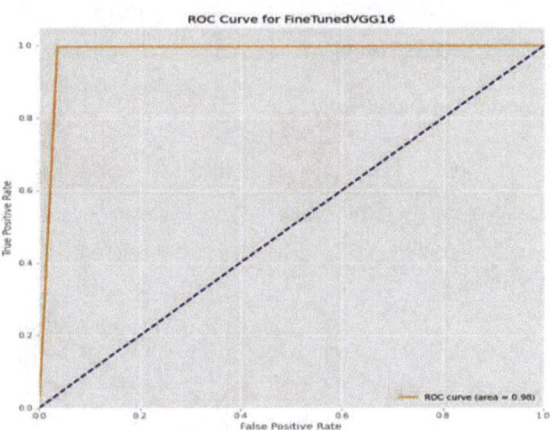

Table 3 Comparative analysis of previous studies of COVID-19 detection

Related studies	Training samples type	Training data size	Model architecture	Accuracy (%)
Wang et al. [34]	X-ray samples(chest)	195 Positive 258 Normal	M-Inception	82.9
Ying et al. [36]	CT images(chest)	777 Positive 708 Normal	DRE- Net	86.0
Xu et al. [37]	CT images(chest)	219 Positive 224 Viral pneumonia 175 Normal	ResNet + Location Attention	86.7
Hemdan et al. [38]	X-ray samples(chest)	25 Positive 25 Normal	COVIDX-Net	90.0
Zheng et al. [35]	Chest CT images	313 Positive 229 Normal	3D Deep Network + UNet	90.8
Wang and Wong [25]	X-ray samples(chest)	53 Positive 5526 Normal	COVID-Net	92.4
Ioannis et al. [39]	X-ray samples(chest)	224 Positive 700 Pneumonia 504 Normal	VGG-19	93.48
Sethy and Behra [26]	X-ray samples(chest)	25 Positive 25 Normal	ResNet50 with SVM Classifier	95.38
Tulin Ozturk et al. [37]	X-ray samples(chest)	125 Positive 500 Normal	DarkCovidNet	98.08
Proposed model	X-ray samples(chest)	141 Positive 1341 Normal	VGG16 + Transfer Learning	99.458

training samples. The model performance can be improved by more samples. In addition, we have implemented an effective screening process for separating the frontal chest X-ray images. The major factors of our approach are as follows:

- The images are gone through aggressive image augmentation techniques to reduce overfitting.
- Normally, the VGG16 model takes about 138 million parameters but by freezing the layers and transfer learning the parameter size is reduced to 14 million parameters which is more time efficient.
- It could act as an effective and precise assistance to the experts.
- The proposed model does not deal with extraction of features which also reduces its complexity.
- Transfer learning and fine-tuning the last fully connected layer perform very well with respect to the VGG16 model alone.
- The results are impressive although the sample space is very low, and also the consistency is maintained along each fold.

In future studies, we will work more on feature extraction and segmentation, which will provide a more accurate analysis on COVID-19. The model can be deployed in the cloud, and it would act as an automated software of COVID-19 diagnosis very fast.

5 Conclusion

COVID-19 pandemic is really a devastating curse for human mankind, and the daily rise of infections and death rates is shocking. In this situation, the application of AI and deep learning plays a vital role to ease this situation. In this paper, the proposed model is an end-to-end architecture which will not need any manual feature extraction methods. It performs aggressive image augmentation to reduce the overfitting issues and also fine-tuned the fully connected layers of the proposed model. It has been implemented for binary COVID-19 disease detection and ready to be tested for larger-scale data sets. In rural areas, where enough testing kits are not available and also for immediate assistance, this proposed automated detection system will fill the shortage of expert radiologists immensely. The only limitation of this model is that it has been tested only in the small sample space. In future, this proposed model will be deployed in the cloud and working on finding more X-ray samples and CT images for comparative analysis of the proposed model.

References

1. World Health Organization, and World Health Organization (WHO). "Pneumonia of unknown cause–China', Emergencies preparedness response." Disease outbreak news (2020)
2. W. Kong, P.P. Agarwal, "Chest imaging appearance of COVID-19 infection." Radiol: Cardiothoracic Imaging. **2**(1), e200028 (2020)
3. https://www.who.int/health-topics/coronavirus 20.03.2020
4. C. Huang, Y. Wang, X. Li, L. Ren, J. Zhao, Y. Hu, L. Zhang, G. Fan, J. Xu, X. Gu, Z. Cheng, T. Yu, J. Xia, Y. Wei, W. Wu, X. Xie, W. Yin, H. Li, M. Liu, Y. Xiao, H. Gao, L. Guo, J. Xie, G. Wang, R. Jiang, Z. Gao, Q. Jin, J. Wang, B. Cao, Clinical features patients infected with 2019 novel coronavirus in Wuhan, China. The Lancet **395**(10223), 497–506 (2020)
5. https://www.worldometers.info/coronavirus/
6. International Committee On Taxonomy of Viruses (ictv) website. https://talk.ictvonline.org/. Accessed 14 Feb 2020
7. E. Mahase, Coronavirus: covid-19 has killed more people than SARS and MERS combined, despite lower case fatality rate. The BMJ. **368**, m641, https://doi.org/10.1136/bmj.m641, 2020
8. https://www.who.int/health-topics
9. T. Ai, Z. Yang, H. Hou, C. Zhan, C. Chen, W. Lv, X. Tao, Z. Sun, L. Xia, Correlation of chest CT and RT-PCR testing in coronavirus disease 2019 (COVID-19) in China: a report of 1014 cases. Radiology **200642**, 1–23 (2019). https://doi.org/10.1148/radiol.2020200642
10. Radiopedia Blog: https://radiopaedia.org/articles/covid-19-3
11. The Hindu Magazine: https://www.thehindu.com/sci-tech/science/covid-19-what-are-the-different-types-of-tests/article31507970.ece
12. L. Lan, D. Xu, G. Ye, C. Xia, S. Wang, Y. Li, H. Xu, Positive RT-PCR test results in patients recovered from COVID-19, Jama. **323**(15), 1502–1503, (2020)
13. Z.Y. Zu, M.D. Jiang, P.P. Xu, W. Chen, Q.Q. Ni, G.M. Lu, L.J. Zhang, Coronavirus disease 2019 (COVID-19): a perspective from China, Radiology (2020), https://doi.org/10.1148/radiol.2020200490 In press
14. W. Kong, P.p. Agarwal, "Chest imaging appearance of COVID-19 infection." Radiol: Cardiothoracic Imaging. **2**(1), e200028
15. https://radiologyassistant.nl/chest/covid-19-ct-findings-in-25-patients
16. G. Litjens, T. Kooi, B.E. Bejnordi, A.A.A. Setio, F. Ciompi, M. Ghafoorian, C.I. Sánchez et al., A survey on deep learning in medical image analysis
17. Y. Chen et al., Deep feature extraction and classification of hyperspectral images based on convolutional neural networks. IEEE Trans. Geosci. Remote Sens. **54**(10), 6232–6251 (2016)
18. P. Rajpurkar et al., "Chexnet: radiologist-level pneumonia detection on chest x-rays with deep learning." arXiv preprint arXiv:1711.05225 (2017)
19. I. Shahzadi et al., "CNN-LSTM: cascaded framework for brain Tumour classification." 2018 IEEE-EMBS Conference on Biomedical Engineering and Sciences (IECBES). IEEE (2018)
20. X. Huang, J. Shan, V. Vaidya, "Lung nodule detection in CT using 3D convolutional neural networks." 2017 IEEE 14th International Symposium on Biomedical Imaging (ISBI 2017). IEEE (2017)
21. F. Caobelli, Artificial intelligence in medical imaging: game over for radiologists? Eur. J. Radiol. **126**, 108940 (2020)
22. Kaiming He, Xiangyu Zhang, Shaoqing Ren, and Jian Sun. Delving deep into rectifiers: Surpassing human-level performance on imagenet classification. In Proceedings of the IEEE international conference on computer vision, pages 1026–1034 (2015)
23. D. Ardila, A.P. Kiraly, S. Bharadwaj, B. Choi, J.J. Reicher, L. Peng, D. Tse, M. Etemadi, W. Ye, G. Corrado et al., End-to-end lung cancer screening with three-dimensional deep learning on low-dose chest computed tomography. Nat. Med. **25**(6), 954–961 (2019)
24. A. Narin, C. Kaya, Z. Pamuk, "Automatic detection of coronavirus disease (covid-19) using x-ray images and deep convolutional neural networks." arXiv preprint arXiv:2003.10849 (2020)
25. L. Wang, A. Wong, "COVID-Net: a tailored deep convolutional neural network design for detection of COVID-19 cases from chest radiography images." arXiv: arXiv-2003 (2020)

26. P.K. Sethy, S.K. Behera, "Detection of coronavirus disease (covid-19) based on deep features." Preprints 2020030300 (2020): 2020

27. T. Ozturk et al., "Automated detection of COVID-19 cases using deep neural networks with X-ray images." Comput. Biol. Med. 103792, (2020)

28. J.P. Cohen, P. Morrison, L. Dao, "COVID-19 image data collection." arXiv preprint arXiv: 2003.11597 (2020)

29. https://www.kaggle.com/paultimothymooney/chest-xray-pneumonia

30. K. Simonyan, A. Zisserman, "Very deep convolutional networks for large-scale image recognition." arXiv preprint arXiv:1409.1556 (2014)

31. F.N. Iandola et al., "SqueezeNet: AlexNet-level accuracy with 50x fewer parameters and < 0.5 MB model size." arXiv preprint arXiv:1602.07360 (2016)

32. M. Hon, N.M. Khan, "Towards Alzheimer's disease classification through transfer learning." 2017 IEEE International conference on bioinformatics and biomedicine (BIBM). IEEE (2017)

33. M. Al Mufti et al., "Automatic target recognition in SAR images: comparison between pre-trained CNNs in a transfer learning based approach." 2018 International Conference on Artificial Intelligence and Big Data (ICAIBD). IEEE (2018)

34. S. Wang et al., "A deep learning algorithm using CT images to screen for Corona Virus Disease (COVID-19)." MedRxiv (2020)

35. C. Zheng et al., "Deep learning-based detection for COVID-19 from chest CT using weak label." medRxiv (2020)

36. Y. Song et al., "Deep learning enables accurate diagnosis of novel coronavirus (COVID-19) with CT images." medRxiv (2020)

37. X. Xu, X. Jiang, C. Ma, "Deep learning system to screen coronavirus disease 2019 pneumonia. arXiv e-prints 2020"

38. E.E.D. Hemdan, M.A. Shouman, M.E. Karar, "Covidx-net: A framework of deep learning classifiers to diagnose covid-19 in x-ray images." arXiv preprint arXiv:2003.11055 (2020)

39. I.D. Apostolopoulos, T.A. Mpesiana, "Covid-19: automatic detection from x-ray images utilizing transfer learning with convolutional neural networks." Phys. Eng. Sci. Med. (2020)

Vehicle Detection and Count in the Captured Stream Video Using Machine Learning

Soumen Kanrar, Sarthak Agrawal, and Anamika Sharma

Abstract The technology of vehicle detection in the captured video has implementation in the variety of fields. This emerging technology when implemented over the real-time video feeds can be beneficial for the missions of search and rescue in remote areas, where access is hampered by mountains and vast land areas without road networks. Areas afflicted by natural disaster (earthquake, flood) may be aided and improved by autonomous UAV systems, in military applications and civilians due to its ability to operate over large, difficult terrain. Rescue or surveillance missions include people and vehicle detection from aerial platforms. The supreme benefit of vehicle detection in the real-time streaming video feed is to track the terrorist intrusion over the borders and hit them. The recent terrorist attack occurred in the Indian soil at Uri, Pulwama, which could be avoided by using the enhanced vehicle detection techniques implied in the live stream video by machine learning. This chapter presents to recognize and identify the moving objects in the live video stream for specific interest. This chapter aims to explore the existing challenging issue in the area of unsupervised surveillance and security. The detection of vehicles is implemented with enhanced algorithms and machine learning libraries like OpenCV, TensorFlow, and others. The various approaches are used to identify and track the specific object through the trained model from the captured video.

Keywords Image classifications · Video tracking · Information gathering · Information analysis · Target detection · Vehicle detection · Vehicle count

S. Kanrar (✉)
Department of Computer Science and Engineering, Amity University Jharkhand, Ranchi 834001, India

Vidyasagar University, Midnapore 721102, West Bengal, India

S. Agrawal · A. Sharma
DIT University, Dehradun, Uttarakhand, India

© The Author(s), under exclusive license to Springer Nature Singapore Pte Ltd. 2021   79
M. Bandyopadhyay et al. (eds.), *Machine Learning Approaches for Urban Computing*,
Studies in Computational Intelligence 968,
https://doi.org/10.1007/978-981-16-0935-0_5

1 Introduction

Objective of this chapter briefly presents the methodology for the detection and count of a particular type of vehicle in the streaming video. This work is required a specific mechanism in the surveillance and security. Efficient identification of a vehicle in live stream video requires massively enhancement in the surveillance techniques to counter the terrorism. In high speed stream handle, the live video stream is also simultaneous copied to the storage server to explore the insights into the chunk of video frames for further process [1–3]. This object detection and identification in the captured video aims to recognize the moving vehicle in the chunk video frames to track the movement of that particular target in whole video stream [4].

Machine learning technologies are successfully used in the face detection and face recognition. The video object co-segmentation is some task of computer vision in which it can be widely used. The specific types of the vehicle count help to enhance the surveillance technique for the captured live stream. It can be done with the help of various machine learning technologies using Python like OpenCV and Tensor-Flow. Computer vision can also be used in object detection, object tracking, object classification, video surveillance, and background modelling [2]. We can consider numerous examples, tracking of a football and the cricket bat in the football and cricket matches, respectively. On the other hand, it is to recognize the people in the stream video or to detect the movement of car, bike, and truck in the stream. That is an example to detect a specific target in surveillance video footage. The vehicle detection in the streaming video also aims to solve some real-time problem, for example, the issue that video feeds cannot be processed in real [5, 6]. We focus on the vehicle detection in the captured video stream so that the track of the total vehicles can be maintained and the total moving vehicles in the captured video stream can be calculated [7]. Further, we can explore the phases like classification of the vehicles based on light and heavy vehicles or two-wheelers and four-wheelers or based on brands. The classification of the moving vehicles can achieve using the artificial neural network, AdaBoost algorithm, and support vector machine.

Classification as well as regression problems can be done with the help of one of the most popular supervised learning algorithms, which is support vector machine [8]. Artificial neural network is a system complex in nature and based on information passed through it can adjust the weights of connection. It is widely used in neural network and data mining concepts [9]. AdaBoost is boosting algorithm, which can be conjugated to improve performance with types of learning algorithm. It is a quite difficult task to keep track of the moving vehicles in the real-time video feed [10]. So it is required to design a methodology to maintain the total count of vehicles in the live video feed. With the help of machine learning libraries like OpenCV, ML library like TensorFlow and other approaches are used to train the model to recognize a particular object. Further, those approaches are to be combined to identify and track the specific object through the trained model. After the recognition of moving vehicles in the captured video stream, we can be further studied to identify the specific vehicle type [11]. The study of the vehicle type can be used for vehicle tracking after the completion of previous phases. Recognize the moving vehicles in the captured video stream and to keep record of the total count of the vehicles. The recognition of the moving vehicles in the captured video stream is an analysis further, which is followed by the identification and tracking of a particular object in the real-time video feed [11]. This aims for the existence challenging issue in the area of surveillance and security. It requires to develop methodology that will detect the specific object in a fast and accurate manner (able to detect a specific object at a speed of nearly 30 frames per second). So that it will be used in the captured video stream likes the real-time video stream. The detection of vehicles has been implemented with the help of ML libraries like OpenCV, and the other approaches are used to train the models to recognize a particular object with the help of ML libraries like TensorFlow. Further approaches are to be combined to identify and track a specific object through the trained model. It involves going through research work on object tracking and trying to develop the best out of it. This chapter mainly focuses on the vehicle detection in the captured video stream, so that we can efficiently track the total vehicles that can be maintained. The total number of vehicles in the captured video stream can be calculated. Further, in the phases, we can include the classification of the vehicles based on light and heavy vehicles or two-wheelers and four-wheelers or based on brands.

2 Overview of the Approach for Vehicle Detection in the Captured Video Stream

The vehicle detection in the captured video stream is done with the help of the OpenCV through which the count of the total number of moving vehicle is being maintained. The initial approach is to detect and count the total number of moving vehicles in the captured video stream with the help of ML libraries like OpenCV. The other approach is to train the model to recognize a particular object with the help of ML libraries like TensorFlow. Further, both approaches are to be combined to identify and track the specific object through the trained model. The approach is used to create a model with the help of two different ML approaches. The initial approach is the implementation of the OpenCV library. With the help of this library, it becomes easier to maintain the record of the moving vehicles in the video stream which is further targeted to be achieved with the help of the training of the model. The basis of the second approach is the implementation about the Inception-ResNetV2. It is one of the fastest algorithms available now. This model was presented in a paper by Google.Inc on August 23, 2016. In 2012, when AlexNet was presented, it was found out that deeper neural networks are needed for further classification. However, processing it takes more time. So, it is trade-off between accuracy and speed. Google.Inc came up with new approach and divides the model in numbers of modules. This approach is also used in the Inception-ResNetV2. It takes the input image of size 299*299*3 in stem cell of the model. It is further divided into two branches and so that the processing can be divided. After that, the results are concatenated. To detect the object, primarily it is considered the Python-based machine learning libraries of TensorFlow, OpenCV, Keras, Pandas, sklearn, Matplotlib, JSON, and PIL. InceptionV4, Inception-ResNet, NumPy, Itertools, and shutil are being used to achieve the target output from this phase.

2.1   Chapter Perspective

Work perspective gives the overview of different kinds of point of view of the approach. The perspective of the work is divided into two kinds of point of views that include developers view and users view. The user view means that the view or perspective which is developed based on the work interface used by the user. The developer's view means the perspective of the work assuming work functionality, control of the developer over the work model.

- **Users View**

As per user point of view, new user can use the model to get the total count of the moving vehicles in the streaming video and also can detect a particular object. In the future, users can use the system in multisector such as defence and security surveillance, gamming, and virtualization. To classify, detect, and track images and video frames made it fast and simpler for ease of use of the users.

- **Developers View**

The developer's point of view is to design a system which can detect the total vehicle's count with the help of OpenCV library. With contrary to vehicle recognition, it is required to develop a model where we could train our model with the help of TensorFlow, for example, to detect the tennis ball from multiple ball images. This concept can be in the future used to detect a particular object in the streaming video.

- **Work Functions**

Work function illustrates the functionality of the working module. The work functionality is to classify images based on trained data sets as per loaded into the system during classifications. The secondary functionality is to recognize the moving vehicle in the captured video stream and to track the movement of that target in whole video. It identifies the total count of the moving vehicles in the captured video.

- **Image Classifications**

The basic work functionality is to classify images on the basis of trained data sets as per loaded into the system during classifications. It can classify images with 78% of accuracy approximately. It is the implementation of Inception-ResNetV2. It is one of the fastest algorithms available now. It takes an input image of size 299*299*3 in stem cell model. It is divided into two branches and so that processing can be divided and then results of them are concatenated. Which can be trained on different data sets of about 1500 images and can be used widely used in different fields also.

2.2   Video Tracking

The secondary functionality is used to track and record the video frames. It labels executions, throughout the object life span. In the current scenario, the model involves the detection of vehicles and records the total count of the moving vehicles in a stable video stream. Real-time computer vision can be achieved with the help of OpenCV library, which is an open-source computer vision library. The OpenCV library includes support vector machine-based algorithms, artificial neural network-based algorithms, K-nearest neighbour algorithm, and many others.

2.2.1   Image Classification

Figure 1 (DFD diagrams) is for the level 0 of classification of image that is depicted. Here, the video is provided as input and with the help of Inception-ResNetV2 algorithm. We get the labelled video as the output.

Figure 2 presents the level 1 classification of the image. Here, the captured video is provided as the input. After that, the pre-processing of video frames is done. These normalized video frames are passed through pre-trained model through which labels are mentioned on the frame using predicted class. Finally, we get a labelled frame as an output.

Figure 3 presents the use case diagram. Here, the user will input the images according to the use case model and will get the labelled image. The actor will use the input video in the use case model and will get output labelled video. In the above Fig. 3, with the help of proposed classification techniques like AdaBoost, artificial neural network, and support vector machine, we can obtain the labelled image or the labelled.

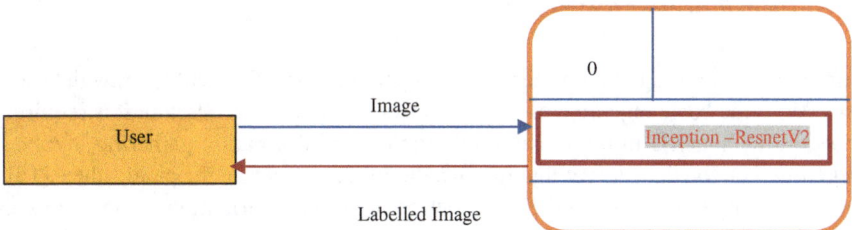

Fig. 1   Visual presentation of level 0

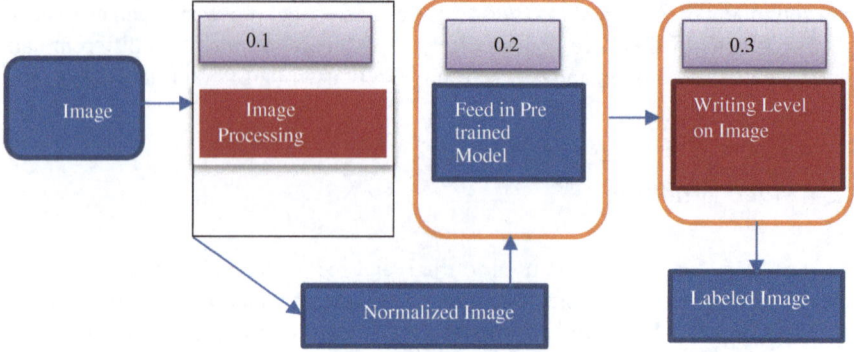

Fig. 2 Visual presentation of level 1

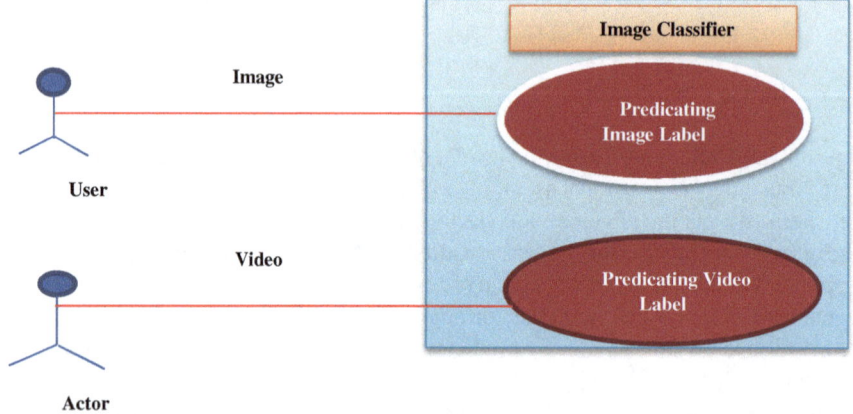

Fig. 3 Visual representation of image classifier

3 Methodology for Vehicle Detection

3.1 *Information Gathering*

Information gathering is the process of collecting information and the raw data set. To achieve the basic objective of vehicle detection in the video stream, it is required to use some analytic in the in captured video. The information gathering involves targeting skeleton data, targeting motion patterns, and targeting the pixel values [12]. Targeting skeleton data means that collection of all the information, which helps to form the basic structure of the model. In vehicle detection, the streaming video we are using, the captured video. We should be prior to know what we have to do with our input feed to get the desired output. We should have prior knowledge how to proceed and what algorithms are to be used. This includes background subtraction; it is a way of eliminating the background from image. To achieve this, we extract the moving foreground from the static background. Next object detection with the help of extraction of features. It is afterwards followed by dimensionality reduction to reduce dependencies of some features. At last, the classification of vehicles is obtained with the help of algorithms like SVM, ANN, and AdaBoost [13]. This information gathering is achieved through targeting skeleton data, targeting motion patterns, and targeting pixel's values.

3.2 Information Analysis

Information analysis is achieved through combining all the information and data gathered. It is used as an input to vehicle detection in streaming video algorithm. The information we collected through the previous information gathering step is combined to achieve the objective of the work. In information analysis, we should understand and implement the various algorithms in the correct manner to be able to detect the moving vehicle in the captured video feed. We should analyse and develop the knowledge of the sequence of implementation of these algorithms. It is required to find out the most suitable values in detection and classification stage. It is important to incorporate the evolutionary optimization techniques in the subject problem. It enhances to find the best symmetry operator for vehicle detection, so the output at every step to be analysed. If the output is not satisfactory, then it is better to develop or search for a better algorithm, which will be accurate and less time consuming in nature. This can help in the process of detection and classification of moving vehicles in the captured video feed.

3.3 Noise Minimization

The collected data should have fewer errors and unwanted inputs. The data inferred should be of very less noise means it should be near to accurate. The noise can be of many types, which can affect the performance of the working algorithms. The noise can be due to the instability of the camera, in the initial stages to detect the moving vehicles. When the filming is started from the camera with the help of UAV in the atmosphere, the rooftops of the buildings and the top of the cars can be confused to be similar in the real-time video streaming [14]. The noise can also occur due to the matching colour of the moving vehicles and the background, which makes it hard for the algorithm to extract the image from the frame of the video with the help of background subtraction.

Noise minimization can be done with the help of dynamic Bayesian network (DBN) reference Fig. 4. In this method, we can incorporate the Bayesian network (BN), which is used for pixel-wise classification. It classifies the pixels of vehicles

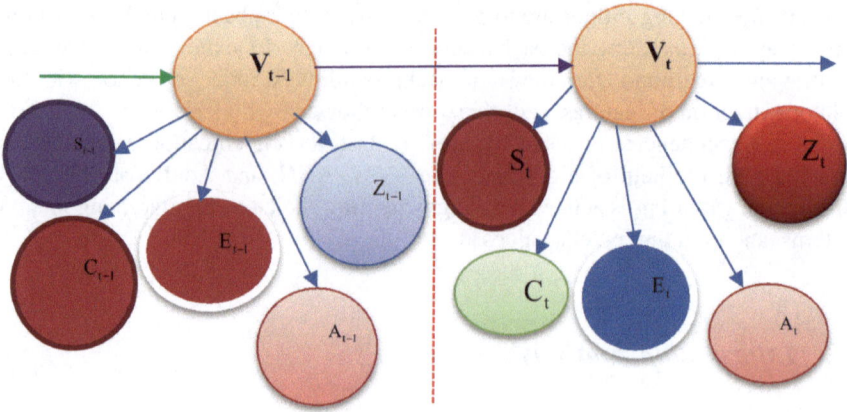

Fig. 4 DBN model

and non-vehicles. Here, V_t means vehicle's pixel at time (t), and it is dependent at the V_{t-1}. For every state of V_t, there is an effect of various factors, which are represented as S_t, A_t, Z_t, C_t, and E_t. Here, S_t is the percentage of pixel. A_t is the aspect ratio. Z_t is the size of vehicle colour connected region. C_t represents the corner feature, and E_t represents an edge feature at time slice (t). Initially, it is required to achieve the conditional probability tables of DBN model. Then, with the help of Bayesian rule, it can be suitable to find the joint probability of the pixel belonging to a vehicle or not at the time slice (t). It is represented below:

$$P(V_t|S_t, C_t, E_t, A_t, Z_t, V_{t-1}) = P(V_t|S_t)P(V_t|C_t)$$
$$P(V_t|E_t)P(V_t|A_t)P(V_t|Z_t)P(V_t|V_{t-1})P(V_{t-1}) \tag{1}$$

Here, $P(V_t|S_t)P(V_t|S_t)$ can be defined as the probability that a pixel is of the vehicle or not. It gives the percentage of the pixel that is classified as colour of the vehicle.

3.4 Designing of Algorithms

Suitable algorithms and programmes are to be developed according to the problem and the collected data. Real-time computer vision can be achieved with the help of OpenCV library, which is open-source computer vision library. OpenCV library includes support vector machine algorithms, artificial neural network algorithms, K-nearest neighbour algorithm, and many others. TensorFlow is also an open-source machine learning platform. With intuitive high-level APIs like Keras building, the training of machine learning models is easy. In the model for detection of a vehicle

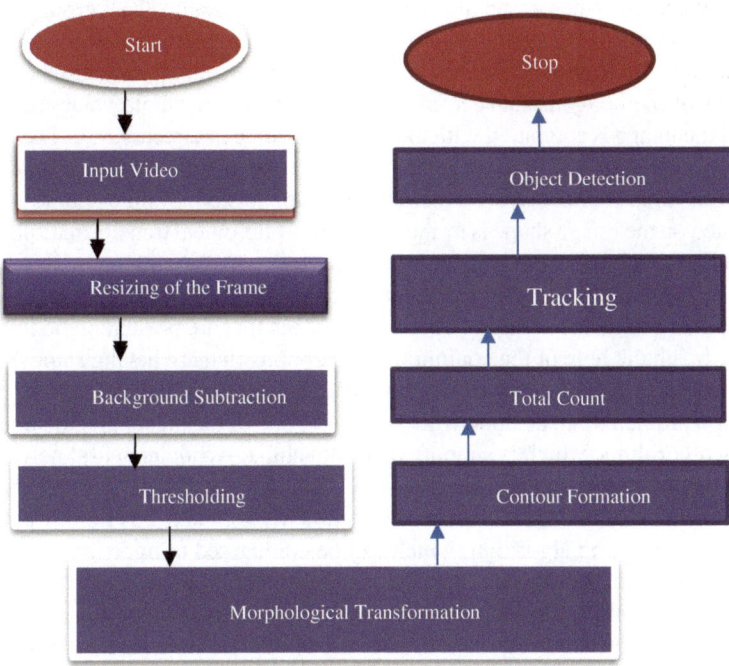

Fig. 5 Framework for the vehicle detection and counting

in moving vehicle, we have to follow several steps to achieve our goal. Below is an example of the steps for the detection of the vehicle in the captured video (Fig. 5).

Algorithm 1: Vehicle Detection and Counting in the video

Step 1: Start

Step 2: Resizing of the video frame

Step 3: Calculating the value of the line to detect vehicles

Step 4: Background Subtraction removes the static background from the foreground

Step 5: Image contrast is set through thresholding

Step 6: Noise is removed with the help of Morphological Transformation

Step 7: Formation of Contours, joining points with same intensity

Step 8: Object detection through contours

Step 9: Tracking of moving vehicle is done

Step 10: Count of the moving vehicles is calculated

Step 11: Stop

From the image when a background is eliminated, it is called background subtraction. The process of background subtraction is the moving foreground it extracted from the static background. Threshold in OpenCV is a technique, where the value of the pixels of the image is set according to the provided threshold value. Each pixel value of the image is compared with the threshold value in thresholding. The value of the pixel of the image is set to zero if its value is less than that of the threshold, otherwise, pixel value is increased to a maximum value, i.e. 255. Simple operation which is operated on the image shape is termed as the morphological transformations. There are two types of inputs required for morphological transformation, one input is the original image, and the second input is termed as the kernel or structuring element. These inputs decide the nature of operation. Shapes that are present in the image are identified with the help of the contours. All the points that are having same intensity are joined using the line around the boundary of image. Classification as well as regression problems can be done with the help of one of the most popular supervised learning algorithms, which is support vector machine. Artificial neural network is a system, complex in nature and based on information passed through it can adjust the weights of connection. It is widely used in neural network and data mining concepts. AdaBoost is boosting algorithm, which can be conjugated to improve performance with types of learning algorithm. For the classification of the vehicles in the captured video, one of the algorithms is used, AdaBoost (feature extraction). For vehicle classification AdaBoost algorithm is very popular, but the training process is quite time consuming. To tackle the limitation of AdaBoost for 'vehicle classification of a rapid learning algorithm' is proposed. Finally, the performance of AdaBoost is improved with the help of the rapid incremental learning algorithm of AdaBoost. Performance of AdaBoost is improved with the help of the rapid incremental learning algorithm of AdaBoost. The following is the incremental learning algorithm with the assumed time complexity $O(n * n)$.

3.5 Target Detection

In the captured stream video for vehicle detection, all the gathered information is required to achieve the objective. The collected information through the previous information gathering step is combined to achieve the objective of the work. The collected data should have fewer errors and unwanted inputs. The data inferred should be of very less noise means it should be near to accurate. Suitable algorithms and programmes will be developed according to the problem and the data that was being collected. Target is detected with the help of algorithms in the captured stream video. After the detection of moving vehicle in the captured stream video, the total count of the moving vehicles is calculated. Then, in the future, the tracking of particular vehicles can be done in the real-time video feed.

Algorithm-2: Incremental Learning of AdaBoos

Input

  (1) O: the original training samples

  (2) n: the number of O

  (3) P: the new samples

  (4) m: the number of P

  (5) Ω: O + P

  (6) Γ: The selected key feature set based on the original training samples

Begin

  (1) Initialize the weights on P

$$u_1(i) \leftarrow 1/m \quad i = 1, 2, \ldots, m$$

  (2) Initialize the weights on Ω:

$$v_1(i) \leftarrow 1/(m+n) \quad i = 1, 2, \ldots, m+n$$

  (3) For $t = 1$ to T

    (1) Normalize the weights on Ω

$$v_t(i) \leftarrow v_t(i)/\sum_{i=1}^{m+n} v_t(i) \quad i = 1, 2, \ldots, m+n$$

    (2) Find new key features on P

      (a) Normalize the weights

$$u_t(i) \leftarrow u_t(i)/\sum_{i=1}^{m} u_t(i) \quad i = 1, 2, \ldots, m$$

      (b) Based on all Haar-like features, Find the weak classifier ϕ_t with the minimum error ε_t

$$\varepsilon_t \leftarrow \tfrac{1}{2}\sum_{i=1}^{m} u_t(i)|\phi_t(\delta_i) - y_i|$$

      where δ_i denotes the value of the selected key Haar-like feature δ on the ith sample

    (3) If $\delta \notin \Gamma$, then $\Gamma \leftarrow \delta$

    (4) Based on Γ, Find the weak classifier f_t which has the minimum error E_t on Ω

$$E_t \leftarrow \tfrac{1}{2}\sum_{i=1}^{m+n} v_t(i)|f_t(\eta_i) - y_i|$$

      where η_i denotes the value of the selected Haar-like feature η on the ith sample

    (5) Compute $\alpha_t \leftarrow \tfrac{1}{2}\ln((1 - E_t)/E_t)$

    (6) Update the weight values on Ω

$$v_{t+1}(i) \leftarrow v_t(i) * e^{\alpha_t *(1 - |f_t(x_i) - y_i|)}$$

    (7) Update the weight values on P

$$u_{t+1}(i) \leftarrow \exp(-y_i\psi_t(x_i)),$$

$$\text{where } \psi_t(x) = \begin{cases} 1 & p_t\delta(x) \leqslant p_t\theta_t \\ -1 & otherwise \end{cases}$$

  End for

End

Output

  (1) AdaBoost classifier: $F(x) = \text{sign}(\sum \alpha_t f_t(x))$

  (2) The key Haar-like feature set Γ

3.6 Module to Identify and Count the Moving Vehicles

The main aim of 'vehicle detection in the streaming video' is to detect any specific vehicles in the captured video stream. This vehicle detection in the captured video aims to recognize the moving vehicle in the captured video stream, to track the movement of that target in whole captured video, and to identify the total count of the moving vehicles in the captured video with the help of various machine learning technologies using Python like OpenCV.

4 Important Terminologies

4.1 OpenCV

Open-source computer vision library (OpenCV) is an open-source computer vision and machine learning software library. OpenCV was built to provide a common infrastructure for computer vision applications and to accelerate the use of machine perception in the commercial products. High-level interface was provided for processing, capturing, and presenting image data. So, for the example, we consider abstracting details on array allocation, camera, and hardware. OpenCV is interoperable and high level and not easy to operate for users. It needs versatility of OpenCV to come at cost of a setup process. It is complex, organizing, and optimizing application code with some uncertainty. Through the gaming sensor like Kinect, web cams, camera and phones are some of the ways through which computer vision can be accessed by the consumers. For depth, cameras pre-compiled version of OpenCV is not supported. Solutions to the requirements in a standardized data format and high-level language interoperable with scientific libraries such as SciPy and NumPy can be provided by the OpenCV's Python bindings. SciPy is used for technical computing and scientific computing. It is a free and open-source Python library. SciPy can implement interpolation, FFT, image processing, integration, optimization, linear algebra, etc. NumPy is used for adding multidimensional and large matrices and arrays, along with high-level mathematical functions to be operated on the arrays. Total count of the moving vehicles is calculated using OpenCV. Real-time computer vision can be achieved with the help of OpenCV library, which is open-source computer vision library. OpenCV library includes support vector machine algorithms, artificial neural network algorithms, K-nearest neighbour algorithm, and many others. First, it is required to resizing of frames in the stream. With the use of creating the background Subtractor MOG2 function, background of the target object is removed. After that, the boundary noise removal and sharp image are achieved. Now, the contrast of the image is set and noise is removed, and finally object detection through contours achieved by joining points having same colours or intensity. Some of the applications include: facial recognition system, gesture recognition, human–computer interaction (HCI), motion understanding, object identification, segmentation and recognition, and motion tracking.

4.2 Background Subtractor

From the image, when a background is eliminated, then it is called background subtraction. Now, for the background subtraction, the moving foreground is extracted from the static background. There are many uses of background subtraction in everyday life like segmentation. Segmentation is the process of partitioning a digital image into multiple segments. Those are sets of pixels, also known as image objects. It counts the number of visitors, security enhancement, pedestrian tracking, and number of vehicles in traffic [15]. Foreground masking is being learnt and identified. There are some problems in background subtraction, since the moving shadows of the vehicles can be classified as the foreground. There are three algorithms used for background subtraction:

- BackgroundSubtractorMOG:
  It is a background and foreground segmentation algorithm.
- BackgroundSubtractorMOG2:
  It is a background and foreground segmentation algorithm. Due to illumination changes, there is better adaptability of varying scenes.
- BackgroundSubtractorGMG:
  It is a combination of per-pixel Bayesian segmentation and statistical background image estimation.

The OpenCV can implement the background subtraction by firstly creating the object and then implement to apply function on the image.

Step 1:   Create object
           fgbg = cv2.createBackgroundSubtractorMOG2(detectShadows = True)
Step 2:   Apply mask for background subtraction
           Fmask = fgbg.apply(image).

4.3 Morphological Transformations

Simple operation which is operated on the image shape is termed as the morphological transformations. There are two types of inputs required for morphological transformation, one input is the original image, and the other input is termed as the kernel or structuring element. These inputs decide the nature of operation. Some of the basic morphological transformations are erosion, dilation, opening, and closing.

Fig. 6 Erosion

- **Erosion**:

  This morphological transformation is basically eroding the boundaries of the foreground image. This erosion can be imagined like the erosion of the soil. If the value of pixels is under the kernel is 1, then the pixel of the original image is considered to be 1, otherwise, it is eroded to 0 (Fig. 6).
  Erosion = cv1.erode(image, k(kernel), iterations = 1)
- **Dilation**: Dilation is just opposite to erosion. If at least one pixel under the kernel is 1, then only the pixel element of the original image is 1. Hence, it results in the increase in the white portion of the foreground or increase in the size of the foreground portion (Fig. 7).
  Dilation = cv1.dilate(image, k(kernel), iteration = 1)
- **Opening**: In opening, both erosion and dilation occurs. Here, first, the process of erosion occurs then the process of dilation occurs. It basically used to remove the noise in the image or the video frame to extract the foreground portion. This noise is in the form of small white noises, and this noise should be removed to achieve the morphological transformation of opening (Fig. 8).
  Opening = cv1.morphologyEx(image, OPEN, k(kernel))
- **Closing**: In closing, both erosion and dilation occurs. Here, first, the process of dilation occurs, then the process of erosion occurs. It is used to remove the noise from the foreground portion where the small holes within the foreground image or the tiny black points are the noise to the video frame or the image (Fig. 9).

Fig. 7 Dilation

Fig. 8 Opening

Fig. 9 Closing

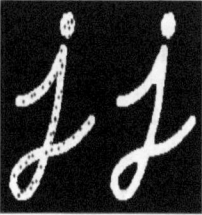

Closing = cv1.morphologyEx(image, CLOSE, k(kernel))

4.4 Thresholding

Threshold in OpenCV is a technique, where the value of the pixels of the image is set according to the provided threshold value. Each pixel value of the image is compared with the threshold value in thresholding. The value of the pixel of the image is set to zero if its value is less than that of the threshold, otherwise, pixel value is increased to a maximum value, i.e. 255. Separation of the foreground portion of the video frame from the background is done with the help of popular segmentation technique of thresholding.

The function (cv. threshold) is used for thresholding in the video frames. A grayscale image is used as the first argument to the source image. Second argument classifies the pixel values, i.e. threshold value. Pixel value exceeding the threshold is assigned with the maximum value which is the third argument. There are many thresholding techniques provided by OpenCV, and it is passed as the fourth argument in the threshold function. There are many types of simple thresholding:

- cv.THRESH_BINARY
- cv.THRESH_BINARY_INV
- cv.THRESH_TRUNC
- cv.THRESH_TOZERO
- cv.THRESH_TOZERO_INV
- ret, imBin = cv2.threshold(fgmask,200,255, cv2.THRESH_BINARY)
- cv.THRESH_BINARY is passed as the parameter in our threshold function. Binary threshold is a simple threshold with values of the pixel, either maximum, i.e. 255 or 0. This can be interpreted as the black portion or white portion. The algorithm 3 presents the steps for thresholding works.
- The black portion or white portion. The algorithm 3 presents the steps for thresholding works (Fig. 10).

Algorithm 3: Discriminant mapping w.r.t Thresholding

Initialization:

f (x, y) = Coordinate Pixel Value

T = Threshold Value.

Compute:

$$\begin{vmatrix} \text{If } f(x, y) > T \\ \text{then } f(x, y) = 0 \\ \text{else} \\ f(x, y) = 255 \end{vmatrix}$$

End

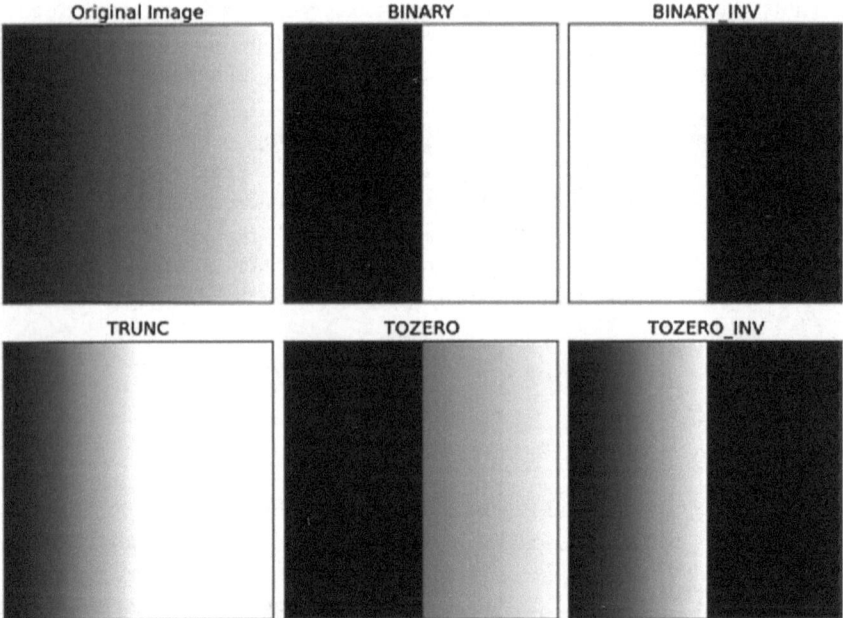

Fig. 10 Thresholding

4.5 Classification of Vehicles

Classification of the vehicles in two-wheeler or four-wheeler is in progress. Classification as well as regression problems can be done with the help of one of the most popular supervised learning algorithms, which is support vector machine. Artificial neural network is a system complex in nature and based on information passed through it can adjust the weights of connection. It is widely used in neural network and data mining concepts. AdaBoost is boosting algorithm, which can be conjugated to improve performance with types of learning algorithm. Classification of the vehicles can be done on the basis of light and heavy vehicles or two-wheelers and four-wheelers or on the basis of brands. Initially, the detected vehicles can be classified separately assuming the type of the two-wheeler or the four-wheeler. After completion of this classification, these vehicles can be further subdivided based on the brands to which they belong. After completion of this classification, these vehicles can be further subdivided on the basis of the brands to which they belong. To keep the track of a particular vehicle, we need to pass the above classification through our training model. After this vehicle identification, track of the vehicle is to be maintained. For the moving vehicles in the random or wrong direction or very close to the camera, there will be some errors while counting the total number of moving vehicles. And for the moving vehicles which pass from the extreme end of the streaming video, the total count of the vehicle also gives error as it is not able to detect the excessive moving vehicles.

The proposed architecture for the vehicle classification with regard to Fig. 11 is as follows:

- Background subtraction and vehicle detection
- Feature extraction
- Dimensionality reduction
- Vehicle classification.

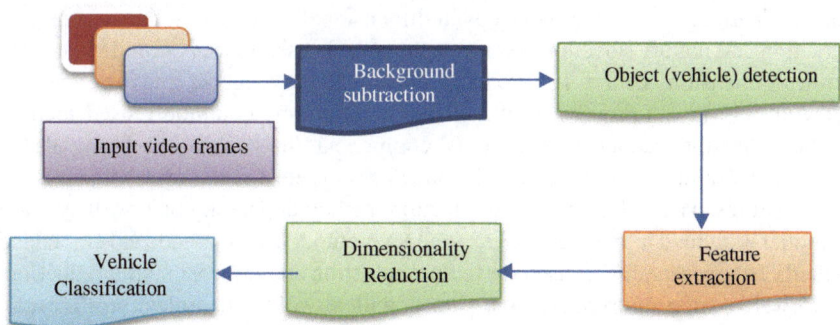

Fig. 11 Proposed architecture for the vehicle classification

4.6 Background Subtraction and Vehicle Detection

From the image when the background is eliminated, then it is called background subtraction. Actually, the background subtraction is the moving foreground, it extracted from the static background. There are many uses of background subtraction in everyday life like segmentation (segmentation is the process of partitioning a digital image into multiple segments. Which are sets of pixels, also known as image objects, counting the number of visitors, security enhancement, pedestrian tracking, number of vehicles in traffic, etc. Foreground masking is being learnt and identified. There are some problems in background subtraction as moving shadows of the vehicles can be classified as the foreground.

4.7 Feature Extraction

In the process of feature extraction in classification of vehicles, there are two ways we can assume to extract the features: appearance based and geometry bases. In appearance-based feature extraction, there is high dimension vector representation of the vehicles [16]. In geometry-based method, we can use the width, area, length, and other's measurement for the vehicles to be detected in the frame. Features that can be considered for classification are: the area of contour, the rectangle around the image contour, ellipse length of axes (major and minor), white pixels between the rectangle around the contour.

4.8 Dimensionality Reduction

Dimensionality reduction is used to reduce the feature, and it is widely used in machine learning. There are two steps in dimensionality reduction that are: feature extraction and feature selection. Feature extraction is used to transform the data into another dimension. Unlike feature extraction, feature selection finds the features that will help in the process of classification. The most commonly used method for dimensionality reduction is principal component analysis (PCA). Using all the features, PCA builds new characteristics and does not select or remove characteristics of the features. The different feature selection method depending on how they build the model is divided into three classes: wrapper method, filter method, and embedded method. Wrapper method is used for feature selection and dimensionality reduction, like forward search algorithm (FSA). Time complexity of feature selection algorithm is $O(n * n)$. Algorithm for feature selection is presented below.

Algorithm 4: Feature Selection

Input: x-features $\in X$

Output: Desired feature set F

1. **Step 1:** Initialization
2. Initialize feature set F = { }
3. **Step 2:** Find X_i with lowest error to F
4. **While** // until select desired number of features
5. For i=1 to feature_count

    Estimate model's error on feature set

    Add: X_i with lowest error to F such that $(F \cup X_i)$
6. End for loop
7. End while loop

---

5   Classification of Vehicles

We can use three different kinds of processes or methods. We can use the methods for the classification of vehicles that are: SVM, AdaBoost, and ANN algorithm.

5.1   ANN Algorithm

The information that is passed can change the internal structure of the complex adaptive system, i.e. artificial neural network (ANN). The weight of the connection is adjusted to achieving it. Weight is associated with each of the connections. Signal between two neurons is controlled by the associated weight, and these weights improve the results by adjusting itself. Artificial neural network is a supervised learning method which means that trainer is superior to the network. The knowledge acquired by artificial neural network is in connected form, in reality, it is difficult to be extracted. Rule for classification is motivated to be extracted through this factor. Data set is the basic component to start the procedure of classification. This data set is a training sample. It is used to learn network, test the sample, and measure the accuracy of the classifiers.

The neural network has the ability to adjust learning ability and network structure by modifying the weight. In the field of artificial intelligence, neural network is very useful. The learning algorithm of the artificial neural network is provided below. The time complexity of the learning algorithm is $O(n * n)$.

Algorithm 5: Learning Algorithm of ANN

Input: Dataset D, Learning rate, Network.

Output: A trained neural network.

Step1: Receive the input.

Step2: Weight the input. Each input sent to network must be weighted i.e. multiplied by some random value between -1 and +1.

Step3: Sum of all the weighted input.

Step4:

Generate output: The output of network is produced by passing that sum through

the activation function.

Output: A trained neural network.

5.2   AdaBoost Algorithm

For the classification of the vehicles in the captured stream video, one of the algorithms used is AdaBoost (feature extraction). For vehicle classification, AdaBoost algorithm is very popular, but the training process is quite time consuming. To tackle the limitation of AdaBoost for vehicle classification, rapid learning algorithm is proposed. To represent the vehicle's appearance, a simple and rotated Haar-like prototypes is introduced. It is used by the algorithm to compute feature pool on a 32 × 32 grey scale image patch. Then, the sample's feature value and the class labels are combined by fast training approach for weak classifiers. Finally, the performance of AdaBoost is improved with the help of a rapid incremental learning algorithm of AdaBoost. AdaBoost algorithm is applied to shallow trees with depth between one and seven. There is nonlinear dependency of complexity.

Algorithm 6: Feature Selection with AdaBoost

Input
  (1) A training set:
    $\{x_i, y_i\}_{i=1}^n, x_i \in X, y_i \in \{-1, +1\}, i = 1, 2, \ldots, n$
    where n is the size of the training set
  (2) x_i denotes the feature vector of the ith sample
  (3) y_i denotes the class label of the ith sample
  (4) X denotes the feature space
Begin
  (1) Initialize weights: $w_1(i) \leftarrow 1/n \quad i = 1, 2, \ldots, n$
  (2) $H \leftarrow null$ // Key feature set
  (3) For $t = 1$ to T
    (1) Normalize the weights:
      $w_t(i) \leftarrow w_t(i)/\sum_{i=1}^n w_t(i) \quad i = 1, 2, \ldots, n$
    (2) For each feature j, train a weak classifier f_j.
    (3) The error ε_j of a classifier f_j is evaluated as follows:
      $\varepsilon_j \leftarrow \sum_{i=1}^n w_{t,i} \kappa(x_i)$
      where $\kappa(x_i) = \begin{cases} 0 & f_j(x_i) = y_i \\ 1 & else \end{cases}$
    (4) Choose the classifier f_t with the lowest error ε_t and $H \leftarrow H \cup \{t\}$
    (5) Compute $\alpha_t \leftarrow \frac{1}{2} \ln((1 - \varepsilon_t)/\varepsilon_t)$
    (6) Update the weights:
      $w_{t+1}(i) \leftarrow w_t(i) * \exp(-\alpha_t f_t(x_i) y_i))$
  End for
  (4) $F(x) \leftarrow sign\left(\sum_{t=1}^T \alpha_t f_t(x)\right)$
End begin
Output
  (1) Key feature set H
  (2) AdaBoost classifier $F(x)$|

Hence, shallow trees have complexity closer to $O(n * n)$. Following is the feature selection with AdaBoost algorithm and the complexity is $O(n * n)$. Performance of AdaBoost is improved with the help of the rapid incremental learning algorithm of AdaBoost. The following is the incremental learning algorithm with the time complexity $O(n * n)$.

Algorithm 7: Incremental Learning of AdaBoost

Input
  (1) A training set:
    $\{x_i, y_i\}_{i=1}^{n}, x_i \in X, y_i \in \{-1, +1\}, i = 1, 2, \ldots, n$
    where n is the size of the training set
  (2) x_i denotes the feature vector of the ith sample
  (3) y_i denotes the class label of the ith sample
  (4) X denotes the feature space
Begin
  (1) Initialize weights: $w_1(i) \leftarrow 1/n$ $i = 1, 2, \ldots, n$
  (2) $H \leftarrow null$ // Key feature set
  (3) For $t = 1$ to T
    (1) Normalize the weights:
      $w_t(i) \leftarrow w_t(i) / \sum_{i=1}^{n} w_t(i)$ $i = 1, 2, \ldots, n$
    (2) For each feature j, train a weak classifier f_j.
    (3) The error ε_j of a classifier f_j is evaluated as follows:
      $\varepsilon_j \leftarrow \sum_{i=1}^{n} w_{t,j} \kappa(x_i)$
      where $\kappa(x_i) = \begin{cases} 0 & f_j(x_i) = y_i \\ 1 & else \end{cases}$
    (4) Choose the classifier f_t with the lowest error ε_t and $H \leftarrow H \cup \{t\}$
    (5) Compute $\alpha_t \leftarrow \frac{1}{2} \ln((1 - \varepsilon_t)/\varepsilon_t)$
    (6) Update the weights:
      $w_{t+1}(i) \leftarrow w_t(i) * \exp(-\alpha_t f_t(x_i) y_i))$
  End for
  (4) $F(x) \leftarrow sign\left(\sum_{t=1}^{T} \alpha_t f_t(x)\right)$
End begin
Output
  (1) Key feature set H
  (2) AdaBoost classifier $F(x)$

Haar-like feature extraction is used in object recognition. A rectangle can be noted with the 5-tuple, $r = (x, y, w, h, a)$, where 'a' is the inclination of the rectangle. $a = \{0, 45°\}$; x, y are the top-left coordinate; w, h are the width and height (Fig. 12).

To encode the local appearance of objects, Haar-like features are used. Two or three connected black-and-white rectangles are consisted by each Haar-like feature. The difference between black-and-white rectangles is the sums of pixel value given the value of Haar-like features. Vehicle image patch is obtained by simple and rotated prototypes.

6 Experiment Results and Discussion

This section presents the vehicle detection in the captured video stream. Here, we gather all the required information to achieve the desire objective. The information gathering involves targeting skeleton data, targeting motion patterns, and targeting the pixel values. The presence of noise due to the instability of the camera, hence, in the initial stages of the work efficiently detects the moving vehicles. Here, the video feed be still with no movement of the camera. The vehicle detection in the streaming

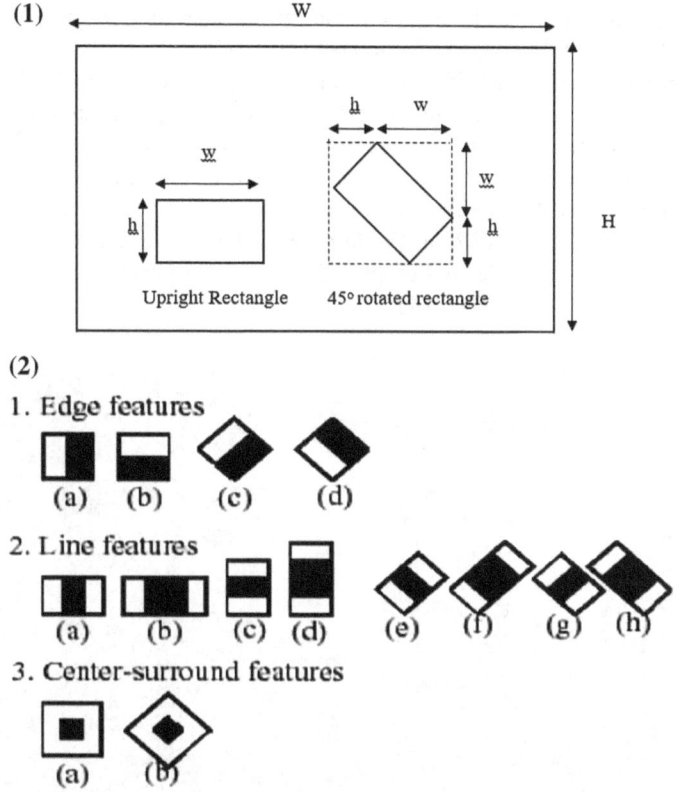

Fig. 12 (1): Two types of Haar-like feature rectangle. (2): Haar-like features

video is captured at the Kishanpur chawk, Mussoorie-diversion road, Dehradun. We have used the tripod to maintain the stability of the camera to get the better results. The output we get from the model that should match the desired results. Validation means to tests the output with the desired output. If the obtained output and wanted output did not match with each other, then we have to modify the used algorithms and check the noise in the captured video. In the model, we first observe the total number of moving vehicles in the captured video through the manual observation. This obser-vation will be the desired result or wanted output. The experimental results are exhibited the identification of the moving object with the help of OpenCV libraries. The total vehicle count is being calculated based on the methodology. Initially, we have to resize the frames of the video stream. With the help of creating background Subtractor MOG2 function, the background of the target object is removed. After that the removal of boundary noise, sharp images are achieved. Now, the contrast of the image is set, and noise is removed. Object detection through the obtained contours is achieved by joining points having same colour or intensity. Then, we move towards the tracking and counting. The experimental results according to the methodology are presented in the following subsection (Fig. 13).

Now, the technique of identification is applied to the moving vehicle. The desired number of detected output detection in the captured video at the daytime should be 27. However, we are received the output as 25. This is less than the desired output. This is happened due to the several reasons, includes vehicles moving at the extreme edge cannot be easily detected, vehicles moving parallel or very close to each other are sometimes considered to be a single object. Hence, we get the experimental results to be less than the desired output. Hence, we get experimental result to be less than the desired output (Fig. 14).

Fig. 13 Experimental result during day time

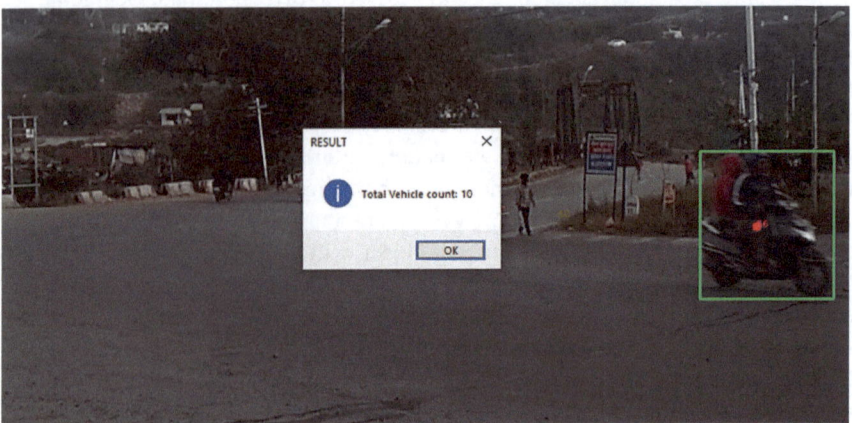

Fig. 14 Experimental result during evening

During the evening, we have conducted the experiment, and the obtained result is 5, but in actual, the desired output should be 9. This video has been captured from the corner of the road not from the middle of the road at the evening time. In this scenario, the experimental result is closer towards the actual result because the light from the car on the ground at the evening time is sometimes detected to be an object. However, some vehicles pass undetected because the video is captured from the side of the road. During the night time, many vehicles passed by but many of them remain undetected. At night, the dispersion of the light is very much; hence, the used algorithm for vehicle detection confuses between the actual moving vehicle and the headlight of that vehicle. Actual moving vehicle and the headlight of that vehicle is considered to be different objects. Even so, at night time due to this confusion, and direct glare of headlight on the camera, this light on the ground cannot be considered to be a different object due to confusion. Hence, the experimental result of vehicle count is less than the desired output. The main aim is to detect specific vehicles in the captured video stream for the vehicle detection methodology. The video detection in the captured video aims to recognize the moving vehicle in the captured video stream, to track the movement of that target in whole video, and to identify the total count of the moving vehicles in the captured video with the help of various machine learning technologies (Fig. 15).

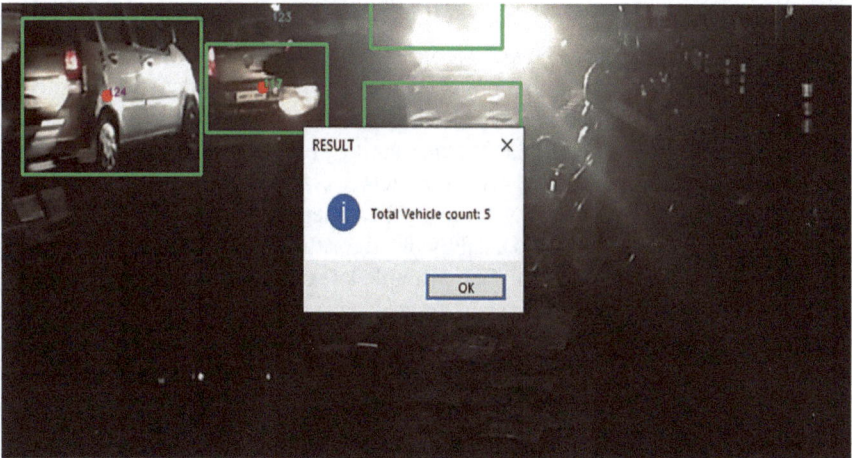

Fig. 15 Experimental result at the night time

Algorithm 8: Accumulation of Haar-like Features

Input: Dataset D, Learning rate, Network.

Output: A trained neural network.

Step1: Receive the input.

Step2: Weight the input. Each input sent to network must be weighted i.e. multiplied by some random value between -1 and +1.

Step3: Sum of all the weighted input.

Step4:

Generate output: The output of network is produced by passing that sum through

the activation function.

Output: A trained neural network.

To encode, the local appearance of objects Haar-like features are used. Two or three connected black-and-white rectangles are consisted by each Haar-like feature. The difference between black-and-white rectangles' sums of pixel value gives the value of Haar-like features. The vehicle image patch is represented by simple and rotated prototypes. For the classification of vehicles in the captured video stream, one of the algorithms used is AdaBoost (feature extraction). For vehicle classification, AdaBoost algorithm is very popular but the training process is quite time consuming. To tackle the limitation of AdaBoost for vehicle classification, rapid learning algorithm is proposed. To represent a vehicle's appearance simple and rotated Haar-like prototypes is introduced, which is used by the algorithm to compute feature pool on a 32 × 32 grey scale image patch. Then, sample's feature value and the class labels are combined by fast training approach for weak classifiers. Finally, the performance of AdaBoost is improved with the help of the rapid incremental learning algorithm of AdaBoost. AdaBoost algorithm is applied to shallow trees with depth between one and seven, and there is nonlinear dependency of complexity. Hence, shallow trees have complexity closer to $O(n \log n)$. The Normal video frame with no application of background subtraction is presented in (Fig. 16).

6.1   Contours Formation

Shape that is present in the image is identified with the help of the contours. All the points that are having same intensity are joined using the line around the boundary of image. The function named findContours in OpenCV is used to identify the contours. This function is used to extract contours from the image and is best used for the binary

Fig. 16 Normal video frame with no application of background subtraction

images. The drawContours function help as to draw the contours. Determining the size of the object of interest, shape, analysis, object detection can be done with the help of contours. The Video frame with application of background subtraction is presented in (Fig. 17). The syntax of both functions is provided below.

Countours0,hierarchy = cv2.findContours(mask,cv2.RETR_EXTERNAL,cv2. CHAIN_APPROX_NONE)
cv.DrawContours (image, contours, contourId, colour, thickness)
area = cv2.contourArea(count) (Figs. 18, 19).

Fig. 17 Video frame with application of background subtraction

Fig. 18 Normal video frame with no application of find contours

Fig. 19 Video frame with application of find contours

6.2 Identification and Vehicle Count

Figures are shown the identification of the moving object with the help of OpenCV libraries. The total vehicle, count is being calculated. First resizing of frames of the streaming video is done. With the use of createBackgrounSubtractorMOG2 function background of the target object is removed. Boundary noise removal and sharp image are achieved. Contrast of the image is set, and noise is removed. Object detection through contours is achieved by joining points having same colour or intensity. Afterwards, the tracking and counting are done (Fig. 20).

This chapter exhibits the methodology to detect any specific vehicles in the captured stream video. The presented vehicle detection methodology for the captured stream, recognizes the movement of the vehicle in the video stream. It is exhibited

Fig. 20 Identification of the moving vehicles in captured video

the total count of the moving vehicles in the captured video with the help of various machine learning technologies. Figure 13 gives the total count of the moving vehicles in the sample video which comes out to be 25. It shows the identification of the moving object with the help of OpenCV. Total count value increases on identifying the moving object, whether it is a two-wheeler or a four-wheeler vehicle or light or heavy vehicle (Fig. 21).

Noise in the video is existed due to the instability of the camera. Hence, in the initial stages of the data collection, ultimate care to be considered that the video feed should be still with no movement of the camera. When filming from the camera with the help of UAV in the atmosphere, the rooftops of the buildings and the top of the

Fig. 21 Total count of the vehicles is determined in the captured video

cars can be confused to be similar in the real-time video streaming. Classification of the vehicles in two-wheeler or four-wheeler is the process to achieve after the identification and counting total moving vehicles in the captured video. Classification as well as regression problems can be done with the help of one of the most popular supervised learning algorithms.

6.3 Non-thermal Surveillance

In the non-thermal surveillance, the dispersion of the light is very much; hence, the algorithm used for vehicle detection gets to confuse between the actual moving vehicle and the headlight of that vehicle. Actual moving vehicle and the headlight of that vehicle is considered to be different objects. To overcome this problem, it is better to use the thermal surveillance technique. The dispersion of light is less at the night. In the day light, it is very feasible to identify and count the total moving vehicles in the streaming video. However, in the night, it becomes very difficult to identify the active vehicles in the streaming video. For this purpose, it is better to use the night-vision camera to identify the moving vehicles in the dim light or the night light (Fig. 22).

6.4 Thermal Surveillance

The thermal camera works by detecting and collecting the radiations form the spectrum of heat that is emitted by the object. In the thermal imaging system, we can give preference to the longer wavelengths than the visible light, and hence we operate

Fig. 22 Non-thermal captured video feed of moving vehicles at night time

Fig. 23 Captured thermal video feed of moving vehicles at night time

Fig. 24 Total vehicle counts in the captured thermal video feed

on the infra-red part of the spectrum. There are differences in the thermic radiation produced by the objects in the images produced by the thermal cameras. In thermal surveillance, detection of vehicles at night is much easier. The dispersion of the light is a low; hence, vehicle detection algorithm can easily identify the moving vehicles in the streaming video. For light vehicles, the dispersion of light is low while the dispersion of light is more for heavy vehicles. The number of the moving vehicles in the sample video comes out 14.

It shows the identification of the moving object with the help of OpenCV. It shows the total count value increases on identifying the moving object, whether it is a two-wheeler or a four-wheeler vehicle or light or heavy vehicle. The video is captured with the help of a thermal camera (Figs. 23 and 24).

7 Scope for Future Work

- **Processing time**

To reduce the time taken by the algorithm to search the frame full, we can search only necessary parts of the image or the input frame. Searching algorithm such as InceptionV4 and Inception-ResNet version 2 can decrease the number of pixels that are to be searched in the image which can make the programme faster.

- **Motion Analysis**

This part of future works will mainly deal with the spreading of the object in the different frames of the video or not. In the present condition, the presented model uses OpenCV for vehicle recognition, furthermore, with the help of TensorFlow and other algorithms. The exhibited model can be further trained to identify and track a particular object.

- **Tracking**

Initially, the vehicles detected can be classified separately assuming the type of the two-wheeler or the four-wheeler. After completion of this classification, these vehicles can be further subdivided based on the brands to which they belong. On the completion of the brand classification, these vehicles can be further classified based on the product type. To keep the track of a particular vehicle, we need to pass the above classification through our training model. After this vehicle identification, track of the vehicle is to be maintained.

- **Arial Supervision**

After the success of the identification and tracking of the particular vehicle or object in the streaming video, the tracking is to be done from the height with the help of the UAV. The particular object remains in the frames of the video due to the tracking movement of the UAV on that specific object or the vehicle. For this, the speed of UAV is to be matched to the speed of the object that is tracked. From height due to various disturbance noise will be captured in the video, which can hinder the tracking process.

8 Conclusion

In this chapter vehicle detection, methodology is presented for object detection for the video surveillance. We have presented our model on InceptionV4 and Inception-ResNet that classify images and label videos based on the training it received. It has many applications in the field of medical diagnoses, defence, games, and virtualization and has high accuracy and high speed compare to other models presented so far. Video detection in the captured video aims to recognize the moving vehicle in the captured video stream, to track the movement of that target in whole video, and to identify the total count of the moving vehicles in the captured video. This is exhibited with the help of OpenCV and TensorFlow. This includes background subtraction; it is a way of eliminating the background from image. To achieve this, we extract the moving foreground from the static background. Afterwards, the thresholding is done to achieve contrast, then noise removal is done with the help of morphological transformation. Object detection is done through contours then finally tracking, and total counting of the moving vehicle is done. Identification of the object is done through

TensorFlow. This work was tested on the tennis ball but can be trained on unusual data sets of about 1500 images and can be used widely used in different fields. There are many problems occurred during the running of our module for vehicle detection. The initial problem is while capturing the video. There should not be any noise. If the video is not still, then the OpenCV cannot identify the moving object precisely and due to whole frame moving, it will detect multiple still object as the moving object. In the night light, the moving vehicles cannot be determined easily in the streaming video. For this purpose, we need to have the night-vision camera for the precise capturing of the moving vehicles in the streaming night light video. In the training of the multiple images in the Google software named Colab took several hours. This process is the time taking process, which gave the best results when trained over the multiple large input images, and we will get good accuracy, e.g. for 1500 images files the accuracy is near to 80%. For the smaller inputs, the accuracy is also tinier. And the same is to be done for the videos also. The optimization of the model is to be done so that the various errors can be corrected to achieve the desired output. We can apply various optimization techniques like feature extraction during detection phase. Regards in this case, we can use the canny edge detector with moment preserving thresholding. Noise minimization can be done with the help of dynamic Bayesian network (DBN). In this proposed method, it is better to use Bayesian network (BN), which is considered for pixel-wise classification. It classifies the pixels of vehicles and non-vehicles.

References

1. S. Kanrar, M. Siraj, "Performance of multirate multicast in distributed network." Int. J. Commun. Netw. Syst. Sci. **3**(6), 554–562, (2010) https://doi.org/10.4236/ijcns.2010.36074
2. S. Kanrar, N.K. Mandal, Video traffic analytics for large scale surveillance. Multimedia Tools Appl. **76**(11), 13315–13342 (2017). https://doi.org/10.1007/s11042-016-3752-0
3. S. Kanrar, N.K. Mandal, Video traffic flow analysis in distributed system during interactive session, Adv. Multimedia. **2016**, Article ID 7829570, pp 1–14 (2016), https://doi.org/10.1155/2016/7829570
4. J. Canny, "A computational approach to edge detection," IEEE Trans. Pattern Anal. Mach. Intell. PAMI **8**(6), 679–698 (1986)
5. Minkyu Cheon, Wonju Lee, Changyong Yoon, Vision-based vehicle detection system with consideration of the detecting location. IEEE Trans. Intell. Transp. Syst. **13**(3), 1243–1252 (2012)
6. B.F. Lin, Y.M. Lin, Fu Li-Chen, P.-Y. Hsiao, L.-A. Chuang, S.-S. Huang, M.-F. Lo, integrating appearance and edge features for sedan vehicle detection in the blind-spot area. IEEE Trans. Intell. Transp. Syst. **13**(2), 737–747 (2012)
7. C.C.R. Wang, J.J.J. Lien, "Automatic vehicle detection using local features—a statistical approach," **9**(1), 83–96 (2008)
8. V. Vapnik, *The Nature of Statistical Learning Theory* (Springer, New York, 1995)
9. J. Han, M. Kamber, J. Pei, "Data mining concepts and techniques", Publisher—Morgan Kaufmann, ISBN 978-0-12-381479-1 (2001)
10. C. Stauffer, W.E.L. Grimson, Learning patterns of activity using real-time tracking. IEEE Trans. Pattern Anal. Mach. Intell. **22**(8), 747–757 (2000)

11. T. Kato, Y. Ninomiya, I. Masaki, Preceding vehicle recognition based on learning from sample images. IEEE Trans. Intell. Transp. Syst. **3**(4), 252–260 (2002)
12. L. Wixson, Detecting salient motion by accumulating directionally-consistent flow. IEEE Trans. Pattern Anal. Mach. Intell. **22**(8), 774–779 (2000)
13. S. Gupte, O. Masoud, R.F.K. Martin, N.P. Papanikolopoulos, Detection and Classification of Vehicles. IEEE Trans. Intell. Transp. Syst. **3**(1), 37–47 (2002)
14. A.C. Shastry, R.A. Schowengerdt, Airborne video registration and traffic-flow parameter estimation. IEEE Trans. Intell. Transp. Syst. **6**(4), 391–405 (2005)
15. S. Kanrar, K. Dawar, A. Pundir, "Pedestrian localisation in the typical indoor environments," Multimedia Tools Appl. (2020), https://doi.org/10.1007/s11042-020-09291-w
16. S. Kanrar, "Dimension compactness in speaker identification," ICIA-16: Proceedings of the International Conference on Informatics and Analytics, Article No: 18, pp, 1–6, (2016), https://doi.org/10.1145/2980258.2980296

Dimensionality Reduction and Classification in Hyperspectral Images Using Deep Learning

Satyajit Swain, Anasua Banerjee, Mainak Bandyopadhyay, and Suresh Chandra Satapathy

Abstract Development in the field of computer-aided learning and testing have stimulated the progress of novel and efficient knowledge-based expert systems that have shown hopeful outcomes in a broad variety of practical applications. In particular, deep learning techniques have been extensively carried out to identify remote sensed data obtained by the instruments of Earth observation. Hyperspectral imaging (HSI) is an evolving area in the study of remotely sensed data due to the huge volume of information found in these images, which enables better classification and processing of the Earth's surface by integrating ample of spatial and spectral features. Nevertheless, because of the high-dimensional data and restricted training samples available, HSI presents some crucial challenges for classification of supervised methods. In particular, it addresses the problems of spectral and spatial resolution, volume of data, and model conversion from multimedia images to hyperspectral data. Various deep learning-based architectures are currently being established to solve these limitations, showing significant results in the analysis of hyperspectral data. In this paper, we deal primarily with the hyperspectral datasets, the dimensionality curse problem, and methods for classifying those datasets using some deep neural networks (DNN), especially convolutional neural networks (CNN). We provide a comparative analysis of various dimensionality reduction (DR) and classification techniques used for finding accuracies based on the datasets used. We also explore certain hyperspectral imaging applications along with some of the research axes.

S. Swain (✉) · A. Banerjee · M. Bandyopadhyay · S. C. Satapathy
School of Computer Engineering, Kalinga Institute of Industrial Technology, Bhubaneswar, Odisha, India
e-mail: swain.satyajit2011@gmail.com

A. Banerjee
e-mail: anasua123.banerjee@gmail.com

M. Bandyopadhyay
e-mail: mainak.bandyopadhyayfcs@kiit.ac.in

S. C. Satapathy
e-mail: suresh.satapathyfcs@kiit.ac.in

© The Author(s), under exclusive license to Springer Nature Singapore Pte Ltd. 2021   113
M. Bandyopadhyay et al. (eds.), *Machine Learning Approaches for Urban Computing*,
Studies in Computational Intelligence 968,
https://doi.org/10.1007/978-981-16-0935-0_6

Keywords Deep learning (DL) · Hyperspectral imaging (HSI) · Deep neural networks (DNN) · Convolutional neural networks (CNN) · Dimensionality reduction (DR)

1 Introduction

Due to many deep learning methods, the computer visualization area has been significantly expanded in the past years. This was primarily because of two reasons: (i) availability of labeled image databases with huge amount of images, and (ii) the computing hardware that made computational speed-up possible [1]. A better way to work with materials, identify or define their properties is to analyze how light interacts with them. This study is known as spectroscopy, which determines the behavior of light in their target and identifies materials on the basis of variable spectral signatures that can be known by the spectrum of the material, or the amount of light in variable wavelengths. It indicates the amount of light that is emitted, reflected, or transmitted from a particular object or target. Imaging spectroscopy works on the way light interacts with the targeted materials. The imaging spectrometers (also known as hyperspectral sensors) typically work in the spectral range of 0.4–2.5 μm, collecting the visible, solar-reflected infrared spectrum from the materials detected. Unlike a digital camera that shoots the target in three colors or bands, a hypespectral sensor measures thousands or hundreds of thousands of spectral wavelengths (Fig. 1). The acquired spectrum is then used to create an image so that every pixel in the image contains the full spectrum.

Hyperspectral imaging provides a 3D data called a data cube. The advanced HSI data areas cover a broad variety of processing methods which can effectively pull out the information found in the hyperspectral cube. The popular techniques amongst these include: (i) spectral unmixing, (ii) resolution enhancement, (iii) image restoration and denoising, (iv) anomaly detection, (v) dimensionality reduction, and

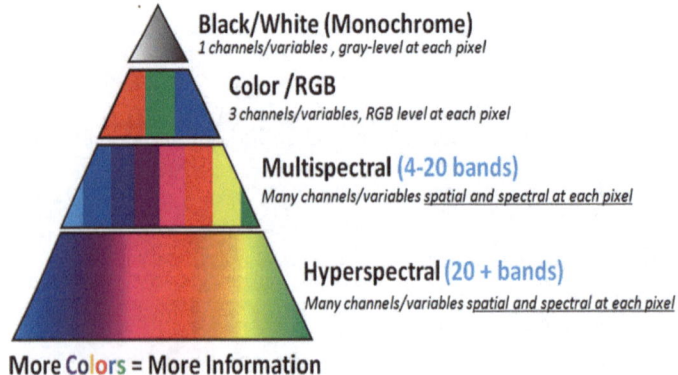

Fig. 1 Hyperspectral wavelengths

(vi) data classification [2]. A broad range of techniques for classifying hyperspectral data depend on machine learning (ML). ML is an area of continuous evolution, however, where new and improved methods are conceived at regular intervals. In this case, new DL models have been developed in the ML field, which have been greatly reinforced by the advancement in knowledge-based computing technologies. These models were a source of inspiration for developing new and improved classifiers for hyperspectral data.

The rest of the chapter is arranged in the following manner. In Sect. 2, the HSI concept in a nutshell is presented with some usable datasets for reference. Some applications of HSI along with its issues and challenges are also discussed in this section. In Sect. 3, various dimensionality reduction components and techniques are presented along with some performance metrics. A comparative analysis of some well-known DR techniques based on the performance metrics is also presented. In Sect. 4, deep learning for classification of hyperspectral images is discussed with the different types of features and layers used, along with the various performance metrics used to evaluate the models. A comparative analysis of various DL models used to find the accuracy using the hyperspectral datasets is also presented, along with some DL limitations. Finally, the paper is concluded emphasizing on the emerging axes of research.

2 Hyperspectral Imaging Analysis

Hyperspectral sensors for a given surface and a given wavelength measure the strength of the radiant flux, capturing the light emitted and reflected by the target as a spectrum of many hundred bands (or channels) per unit of surface (corresponding to one pixel of the image), which defines a curve of spectral response. In action, HSI are tensors (x, y, B), 3D cubes consisting of two spatial dimensions (height x and width y) and one spectral one (with B bands) [3]. This hypercube (Fig. 2) is anisotropic compared to volumetric seismic data cubes, since the three dimensions

Fig. 2 3D projection of a hyperspectral cube

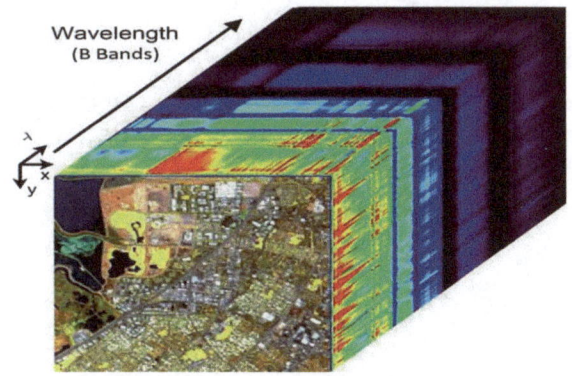

do not reflect the similar physical displacement. This property falls into play when it comes to hypercube conversions and filtering operations.

2.1 HSI Sensors

Over the past decades, many efforts have been made to generate high-quality hyper-spectral data, with the development of a different variety of HSI sensors mounted on either aerial or satellite platforms [2]. Such imaging spectrometers integrate the power of spectroscopy and digital imaging to obtain the corresponding spectral signature for each position in an image plane, using several hundred continuous and narrow bands, obtaining full hyperspectral data cubes by rasterizing the scene covering wide areas of observation. As a consequence, the region or scene being captured is registered in different wavebands, producing a huge data cube. These sensors gather information as image sets in which every image constitutes a spectral band of narrow electromag-netic spectrum wavelength range. Hyperspectral cubes are generally obtained through the airborne sensors such as the Airborne Visible/Infrared Imaging Spectrometer (AVIRIS) from NASA, or by satellites such as the EO-1.

Due to the similarity of the hyperspectral sensor within the spectrum, a prob-able limitation is that the continuous bands tend to be dependent and associated [4]. Situation like this generates redundant information in some bands. However, these sensors create a high-dimensional data space that poses new challenges to the tradi-tional techniques for analysis of spectral data. Table 1 enlists some popularly used hyperspectral sensors, underlining their various spatial and spectral features, such as number of channels, range (in μm), spectral resolution (nm), and the spatial ground sample distance (GSD).

Table 1 Some widely known HSI sensors [2]

Type	Sensor	Bands	Range	Width	GSD
Airborne	AVIRIS	224	0.36–2.45	10	20
	CASI	144	0.36–1.05	2.4	2.5
	HYDICE	210	0.4–2.5	10.2	1–7
	HYMAP	126	0.4–2.5	15	5
	PRISM	248	0.35–1.05	3.5	2.5
	ROSIS	115	0.43–0.86	4	1.3
Satellite	EnMAP	228	0.42–2.4	5.25–12.5	30
	DESIS	180	0.4–1	3.3	30
	HYPERION	220	0.4–2.5	10	30
	PRISMA	237	0.4–2.5	≤12	30
	SHALOM	241	0.4–2.5	10	10

There are generally four methods for sampling the hyperspectral cube (Fig. 3) which are described below.

(i)   **Spatial scanning**: Each 2D output in spatial scanning constitutes a complete slit spectrum (x, λ). In this scanning, the hyperspectral systems create a slit spectrum by projecting a scene strip on the slit and scattering the signal with the help of grating. Such devices pose the limitation of examining the image line by line (using a push-broom scanner), with some mechanical parts inserted in the image optical train.

(ii)   **Spectral scanning**: Every 2D output in spectral scanning constitutes a monochromatic, spatial (x, y) map of the scene. These devices often use optical band-pass filters, either adjustable or a fixed one. The scene is spectrally reflected by changing the filter one after other, while keeping the platform fixed.

(iii) **Non-scanning**: All the spatial (x, y) and spectral (λ) data in non-scanning technique are contained in a single 2D sensor output. Non-scanning HSI tools provide the complete data cube without scanning. These systems provide the benefit of greater light throughput with minimal acquisition time. A 3D structure can be thus obtained from a single snapshot representing a perspective projection of the hypercube.

(iv)   **Spatio-spectral scanning**: Every 2D output in spatio-spectral scanning reflects a spatial scene map with a coded wavelength $(\lambda = \lambda(y))$. It uses a camera (slit + dispersive element) positioned at a nonzero distance behind a

Fig. 3 Hyperspectral data acquisition

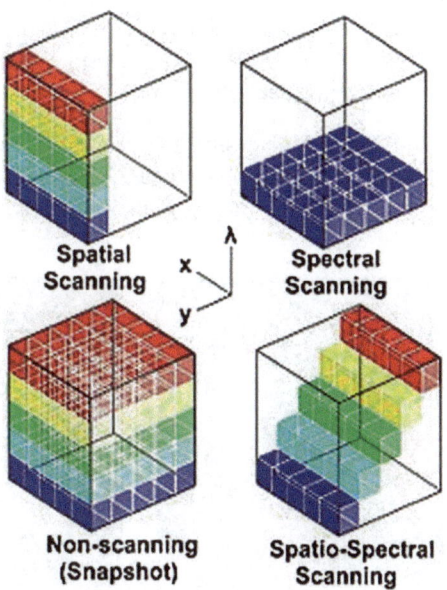

simple slit spectroscope [5]. Enhanced scanning systems can be achieved by placing a dispersive material in front of a spatial scanning system.

2.2 Reference Datasets

Hyperspectral datasets are extensively used for classification, both in the fields of engineering and medical sciences. Medical hyperspectral images are often restricted by low contrast, higher noise, non-homogeneous intensity, which complicate the classification task considerably [6]. Additionally, they tend to face much difficulty in recovery of their shape compared to other types of images. Many public datasets have been made available so far, collected using a hyperspectral sensor. Three different datasets were chosen for our purpose: (i) Indian Pines (IP), (ii) Pavia University (PU), and (iii) Salinas Valley (SV) (Fig. 4). Other datasets include the Kennedy Space Center (KSC), comprising of 13 classes including wetlands and different varieties of wild vegetation, and the Botswana dataset having 14 classes of vegetation and swamps [7]. While features of all are compared in Table 2, the selected ones are presented in details.

(i) **Indian Pines**: IP dataset is recorded via aerial mode with the help of AVIRIS sensor. The plot includes the agricultural areas in Northwest Indiana, USA, with an image of 145 × 145 px having 224 spectral channels. While the majority of the plot contains agricultural lands with different crops, the remainder comprises of dense forests and vegetation. Sixteen classes are labeled, some being very sparse (fewer than hundred samples of oats or alfalfa). Usually, the water absorption bands are separated prior to processing resulting into an image with 200 bands [9]. Despite its small size, it is used as one of the principal reference datasets.

(ii) **Pavia University**: PU dataset is recorded via aerial mode with the help of ROSIS sensor over the Pavia City, Italy. It is split into two parts: Pavia University (610 × 340 px with 103 bands), and Pavia Center (1096 × 715 px with

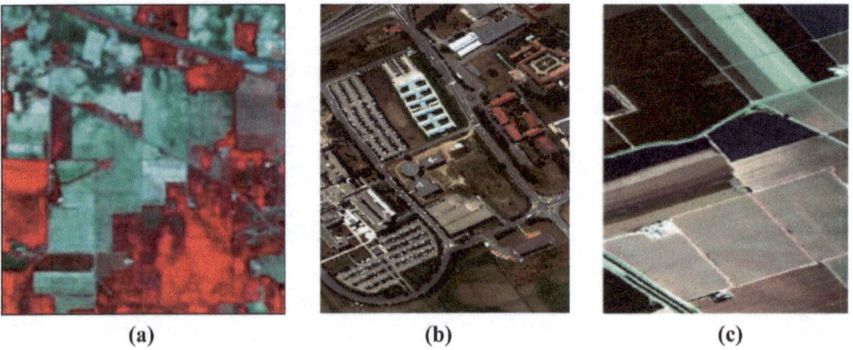

(a)   (b)   (c)

Fig. 4 Natural composite image: **a** Indian Pines, **b** Pavia University, **c** Salinas

Table 2 Public labeled HSI datasets [7, 8]

Dataset ⟶	Indian pines	Pavia university	Salinas valley	KSC	Botswana
Place	Northwestern Indiana, USA	Pavia, Northern Italy	California, USA	Florida	Okavango Delta, Botswana
Pixels	21,025	207,400	111,104	314,368	377,856
Bands (channels)	200	103	224	176	145
Frequency range(μm)	0.4–2.5	0.43–0.85	0.4–2.5	0.4–2.5	0.4–2.5
GSD	20 m	1.3 m	3.7 m	18 m	30 m
Labels	10,249	50,232	54,129	5,211	3,248
Classes	16	9	16	13	14
Mode sensor	Aerial (AVIRIS)	Aerial (ROSIS)	Aerial (AVIRIS)	Aerial (AVIRIS)	Satellite (HYPERION)

102bands) [10]. Nine classes are labeled nearly covering half of the total area. It comprises of different urban materials, vegetation, and water. It is one of the largest labeled hyperspectral datasets used for potentially beneficial applications. Nevertheless, some preprocessing might be required to eliminate certain pixels that do not pose any spectral details.

(iii) **Salinas Valley**: SV dataset is recorded with a low altitude via aerial mode with the help of AVIRIS sensor over the pacific region in the state of California, USA. The plot contains radiance data with 224 bands of size 512 × 217 px, resulting into 16 classes of interest mostly representing various crop varieties. Twenty water absorption bands were discarded before being used for classification.

2.3 HSI Issues and Challenges

HSI data, coupled with their high dimensionality, poses many limitations which makes the task of classification difficult. HSI data suffers from high intraclass variability, leading to unregulated changes in spectrometer-captured reflectance (normally due to changes in atmospheric conditions). The instrumental noise generated by the spectrometer can deteriorate the process of acquisition of data, corrupt the spectral channels to varying degrees, or might render many channels unusable because of calibration or saturation errors. HSI instruments tend to have substantial repetition across adjoining spectral channels, leading to the existence of redundant data which can impede algorithm computational performance.

With images having low or medium spatial resolution, the pixels in the hyperspectral data enfold greater spatial regions on the Earth's surface. As a result, they appear to produce mixed spectral signatures, having higher interclass similarity in boundary areas [2]. Airborne sensors often enfold quite smaller areas than the ones assigned on satellite platforms, resulting in very minimal amount of HSI datasets. Furthermore, the job of marking every pixel found in the hyperspectral dataset is difficult and time taking, since it usually involves human expertise, additionally reducing the number of available datasets for classification purposes.

2.4 Applications of HSI

Remotely sensed HSI is used for a broad variety of tasks in many fields. It is used in the field of agriculture to track crop production and health under environmental stress, related diseases, crop volatility, soil erosion stages, and agricultural precision. In the field of eye care, research is under progress for testing the hyperspectral photography in the diagnosis of macular edema and retinopathy and is tested, preventing any damage to the eye [2]. HSI combined with intelligent software in the food processing industry allows automated sorters to detect and eliminate the defects and foreign content unidentified by conventional camera and laser sorters. With HSI, the geological sample can be easily mapped along with soil composition analysis for almost all minerals of commercial interest. Many minerals such as calcite, feldspar, garmet, olivine, and silica can be recognized from airborne photos.

Hyperspectral monitoring and scanning technology are being applied for various monitoring purposes. HSI is especially convenient in military surveillance because of the countermeasures being taken by military entities to avoid airborne surveillance. Several studies have focused on water quality research, precipitation, and identification of sea ice in water and marine resource management. Soldiers face various chemical hazards, most of them being invisible but can be identified through HSI technology. HSI is used in forest management and environmental management to analyze forest status and health, invasive species detection, and forest plantation infestation [5]. It is also used for tracking pollution from coal and oil-fired power plants, municipal and perilous waste incinerators, cement plants, and various other industrial sources.

3 Dimensionality Reduction in HSI

It can be difficult to imagine a 3D classification problem, while a two-dimensional problem can be drawn in a normal 2D space, and a one-dimensional problem to just a line. There are several feature variables in the ML classification problems on the core of which the predictive classification is made. The more features the training set gets, the harder it gets to imagine and then focus on. Some of these features often become linked, and thus are redundant. This is the case where problems and algorithms related to dimensionality reduction (DR) come into existence. Dimension reduction refers to the method of minimizing the amount of arbitrary variables, by transforming data from a higher dimensional data space to a lower one, generating a set of main variables, and retaining some intrinsic properties of the original data. It is a way to reduce a model's complexity and prevent overfitting, i.e., curse of dimensionality [11].

As discussed in the previous section, high-dimensional data space poses a new challenge to traditional techniques for spectral data analysis. High-dimensional data spaces have been shown to pose the following features: (i) the volume of a hypercube concentrate in the corners, and (ii) the volume of a hypersphere concentrate in an external layer [12]. Most of the hyperspectral data space is empty with minimal training data. There exist two main solutions here: (i) providing larger training datasets, or (ii) reducing dimensionality by drawing out relevant characteristics from hyperspectral signatures. If the first solution is considered, it must be considered that with the increase in number of dimensions, the sample size of the training data will also exponentially increase such that multivariate statistics have accurate estimates. When a hyperspectral sensor is used with several hundreds of spectral channels, consider the impracticality of this example. Thus, the second solution must also be taken into account, having a requirement for feature extraction techniques which can minimize the dimensions of data space without missing the actual information about class separation.

3.1 Components of Dimensionality Reduction

There exist two main components of dimension reduction: (i) feature selection and (ii) feature extraction.

(i)   **Feature Selection**: ML operates on a basic rule—if garbage is put in, all we get is garbage as output, where garbage refers to noisy data. If the number of features is very large this becomes much more significant. We do not need to use every function that is accessible to us to create an algorithm. We can train the algorithm by feeding only those features which really matter. In feature selection, a subset of the actual set of features or variables is obtained that can be used to model the problem. When creating a predictive model, it is the method of raising the number of input variables. There are some good objectives of using feature selection such as it reduces overfitting of data and allows the ML algorithm to be trained faster. It makes the model simple to be interpreted by reducing its complexity. Subsequently, by choosing the right subset, it also improves the accuracy of the model. Feature selection usually involves three ways: filter, wrapper, and embedded.

   (a)   *Filter*: Filter techniques are commonly required in the preprocessing stage. The selection of features does not depend of any ML algorithm. However, features are chosen based on their performance in different statistical tests for their correlation with the expected variable. Linear discriminant analysis (LDA) technique is used to classify a linear combination of features, characterizing, or separating various groups of a categorical variable. The steps of the Filter method is shown in Fig. 5.

(b)   *Wrapper*: A subset of features is generated in wrapper methods, and a model is trained using those features. Addition or deletion of features from the subset is decided on the basis of the conclusion drawn from the preceding model. In essence the problem is minimized to a search space and the techniques are typically very costly in computational terms. Some familiar instances of this method include forward feature selection, backward feature elimination, recursive feature elimination, etc. The steps of the Wrapper method is shown in Fig. 6.

(c)   *Embedded*: Embedded approaches incorporate both the qualities of filter and wrapper techniques. They are used by algorithms which have their own inbuilt methods of selecting the features. LASSO and RIDGE regression are two of the most common examples of such approaches that have built-in penalization functions to minimize overfitting. It works in the following manner as shown in Fig. 7.

(ii) **Feature Extraction**: Feature extraction methods select and/or combine variables into features, significantly reducing the amount of data that must be processed, while still describing the entire dataset accurately. This minimizes

Fig. 5 Filter method

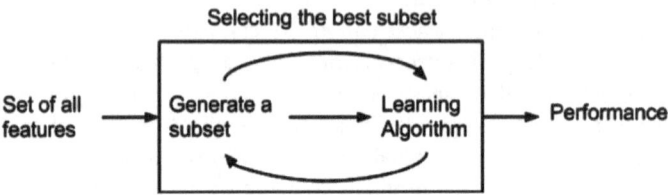

Fig. 6 Wrapper method

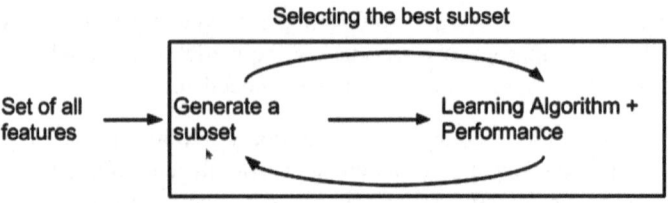

Fig. 7 Embedded methods

the data from a higher dimensional data space to a lower one. It is a dimension reduction method in which a primary collection of raw data is minimized to further reasonable groups for processing. It begins from a primary set of calculated data and builds derived values (features) that tend to be insightful and non-redundant, enabling the successive steps of learning and generalization, and enabling better human interpretations in some cases.

3.2 Techniques for Dimensionality Reduction

Linear techniques, for instance, principal components analysis (PCA) have been used for dimensionality reduction traditionally [13]. A growing number of nonlinear methods for dimension reduction have also been suggested in the past decade [14]. Unlike conventional linear methods, the nonlinear methods are able to handle complex nonlinear data. Some of the DR techniques are discussed below.

(i) **Linear techniques**: Linear techniques for reducing the dimension, embed the data into a lower-dimensionality linear subspace. Some common linear techniques include: (a) principal component analysis (PCA) and (b) linear local tangent space alignment (LLTSA).

   (a) *Principal Component Analysis*: PCA, being an unsupervised linear transformation method, embeds the data into a linear subspace of lesser dimensions. Mathematically, it tries to figure out a linear mapping X that inflates the covariance matrix $cov(Y)$. This linear mapping is obtained by the p principal eigenvectors, i.e. principal components (Fig. 8) of the covariance matrix with the mean zero data [15]. It inflates $cov(Y)$ according to X, with the condition that $|X| = 1$. In a nutshell, PCA tries to figure out the paths of greatest variation in higher dimensional data, and project it into a new subspace of similar or smaller dimensions than the actual one. The new subspace's orthogonal axes (or main components) can be interpreted as greatest variance directions, with the restriction that the new functional axes are orthogonal. While this technique removes the

Fig. 8 Principal components of PCA

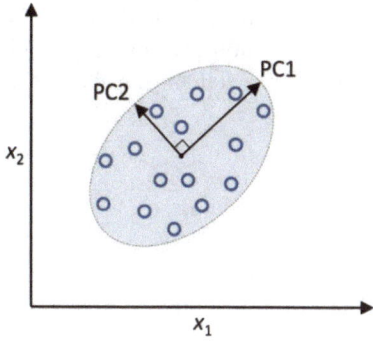

correlated features, the independent variables become less interpretable, and leading to information loss.

(ii) **Nonlinear techniques**: Nonlinear techniques do not depend on the linearity supposition, as a consequence of which large complex data embedding can be defined in the higher dimensional data space. The problem of nonlinear DR can be stated as follows. Consider a dataset constituted in a ($m \times N$) matrix M, comprising of m vectors y_i, $i = \{1,2,...,m\}$ with dimensionality Z [15]. Also, consider this dataset having an inherent dimensionality z, where $z < Z$. In this case, inherent dimensionality implies that the points in dataset D lie within a manifold with dimensionality z inserted in the Z dimension space. DR methods convert dataset X into a new dataset with z dimensions, preserving the geometrical shape of the data to a large extent. Some common nonlinear techniques include kernel PCA (KPCA), independent component analysis (ICA), linear discriminant analysis (LDA), ISOMAP, multi-dimensional scaling (MDS), stochastic neighbor embedding (SNE), etc. Some of these techniques are discussed below.

(a) **Kernel PCA**: KPCA is an extension of PCA applying kernel techniques in the field of multivariate statistics [16]. The original linear functions of PCA are carried out using a kernel with a nonlinear mapping in a reproducing Hilbert kernel space. Because KPCA is a kernel-based technique, its mapping depends on choosing the kernel function K, the polynomial and Gaussian kernel. The kernel is a measure of closeness, i.e., it is equal to 1 when the points coincide, and 0 at infinity. In use, large datasets lead to a large K value, and the problem of storing K may arise. One way to address this issue is to cluster the dataset and fill the kernel with the mean of those clusters. Since this method even can yield a fairly large K, it is suggested to compute the top P eigenvalues and eigenvectors.

(b) **Independent Component Analysis**: ICA describes a generative model for isolating the observed multivariate data into additive subcomponents that is expected to be combination of some undetermined hidden variables. It is a statistical technique used to disclose latent factors underlying the sets of arbitrary variables or signals. It was initially designed for solving the blind source separation (BSS) problem. BSS represents the task of identifying the actual source signals from available combinations, without initial understanding of the mixing techniques. ICA uses a statistical technique as follows [2]:

$$I = X * S \tag{1}$$

where,

$$I = [I_1, I_2 \ldots I_J]^T \tag{2}$$

$$S = [S_1, S_2 \ldots S_K]^T \tag{3}$$

The problem can be defined as follows: Given J linear mixtures of I, of K independent sources S, mixed using an undetermined $M \times N$ mixing matrix X, the principal sources can be estimated from the mixtures [17].

Kernel ICA: Kernel ICA is an extension to ICA that relies on a kernel measure of independence [4]. The method is related to maximizing the independence as reducing the correlation with kernel. Kernel-based learning algorithms use the technique of nonlinear mapping. It relies on canonical correlations in a reproducing kernel Hilbert space (RKHS) R with $k(i, j)$ and feature map $\Phi(i) = k(., i)$ [18]. The R-correlation is denoted as the largest correlation among the two arbitrary variables $a(i)$ and $b(j)$, where a and b lie in the range of R.

$$\rho_R = \max_{a,b \in R} corr(a(i), b(j))$$
$$\text{or,} \quad \rho_R = \max_{a,b \in R} \frac{cov(a(i),b(j))}{\sqrt{\{vara(i)\}\{varb(i)\}}} \tag{4}$$

There are many other kernel functions, such as the Gaussian kernel, which is given by:

$$k(i,j) = \exp\left(-\frac{1}{\delta^2}\|j - i\|^2\right) \tag{5}$$

KICA, being a nonlinear technique for drawing out independent components of images, is able to obtain a notable development in brain MRI images, and the KICA-based classification technique is also useful to classify brain tissues effectively.

FastICA: Like other ICA algorithms, the FastICA method looks for an orthogonal rotation of prewhitened data via a stationary iteration scheme that inflates the estimate of non-Gaussian rotated components. It uses a gradient descent approach to enhance the non-Gaussianity measure, thus stimulating the approximation of the independent components of the source signals. The various dimensionality reduction techniques is shown in Fig. 9.

3.3 Performance Metrics and Quality Criteria

For the evaluation of linear and nonlinear dimensionality reduction techniques, we generally focus on the various performance metrics and quality criterion. The hyperspectral images are represented as a 3D matrix $M(x, y, \lambda)$, with x being the location of the pixel, y as the line number, and λ as the spectral channel. n_x, n_y, n_λ are the number of pixels by line, the number of lines, and the number of spectral channels, respectively [15]. The various performance parameters used are described below.

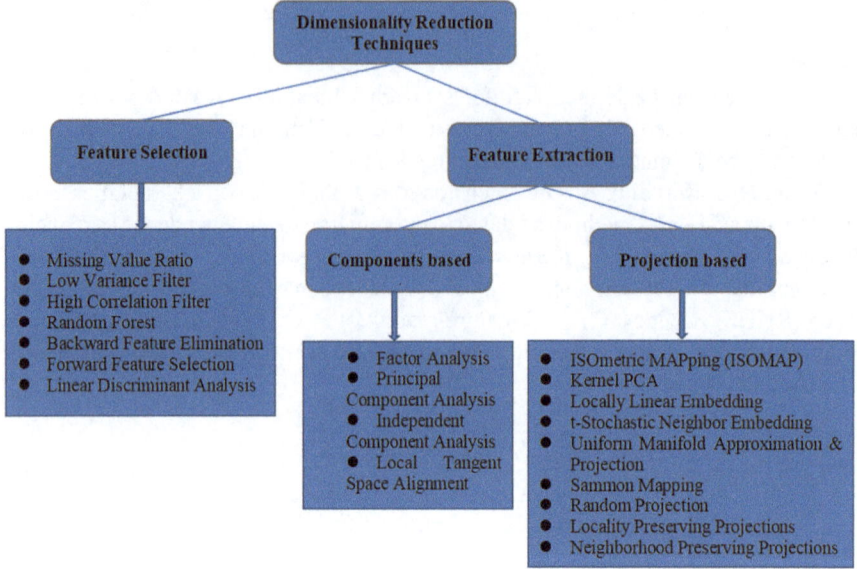

Fig. 9 Taxonomy of dimensionality reduction techniques

(i)   **Mean Absolute Error (MAE)**: It gives the average of the absolute values of
      the errors. Here, it refers to the number of pixels rendered for the measurement,
      irrespective of the image size.

$$\text{MAE}\big(M,\overline{M}\big) = \frac{1}{n_x n_y n_\lambda} \sum_{x,y,\lambda} \big|M(x, y, \lambda) - \overline{M}(x, y, \lambda)\big| \qquad (6)$$

(ii) **Mean Squared Error (MSE)**: It gives the mean of the squared error and is
      more focused on large errors. Here, it also refers to the number of pixels.

$$\text{MAE}\big(M,\overline{M}\big) = \frac{1}{n_x n_y n_\lambda} \sum_{x,y,\lambda} \big|M(x, y, \lambda) - \overline{M}(x, y, \lambda)\big|^2 \qquad (7)$$

(iii) **Spectral Derivative**: In a hyperspectral sensor, the bandwidth of each band
      is a variable. This approach examines the bandwidth variable as a function
      of extra information. It is obvious that if two adjoining bands do not vary
      significantly then only one band can characterize the underlying geospatial
      property [5].

$$\text{SD}(\lambda_i) = \sum_{x} \|M(x, \lambda_i) - M(x, \lambda_{i+1})\| \qquad (8)$$

where M represents the hyperspectral value, x is the spatial position, and λ
is the central bandwidth. Thus, one of the bands is redundant if SD is equal

to zero. Yet adjoining channels that vary greatly should be preserved, while identical ones can be minimized.

(iv) **Entropy**: It is the average level of information or uncertainty inherent in the possible outcomes of a variable. It is basically the measurement of homogeneity of a given dataset. The Shannon information entropy given by Claude Shannon, refers to the amount of data stored or conveyed by a data source. It is given by,

$$Q_i = -\log_2 p_i \qquad (9)$$

where p_i is the probability of the emergence of level i. As the probability p_i ranges between zero to one, the amount of information ranges from zero to infinity. One thing to notice here is that more rare the level is, more big is the amount of information.

(v) **Variance Ratio**: It tells whether the variance of the populations from which the samples have been taken. In the context of dimensionality reduction, it gives us the ratio how populated each sample is based on the number of bands selected.

(vi) **Similarity (SS)**: The similarity criterion measures the similarity between two vectors v and v' of n_λ dimensions. It is given by,

$$SS(v', v) = \sqrt{RMSE(v, v')^2 + \left(1 - corr(v, v')^2\right)^2} \qquad (10)$$

with root mean squared error,

$$RMSE = \sqrt{\frac{\sum_{\lambda=1}^{n_\lambda} (v(\lambda) - v'(\lambda))^2}{n_\lambda}} \qquad (11)$$

and correlation,

$$Corr(v, v') = \frac{\frac{1}{n_\lambda-1} \sum_{\lambda=1}^{n_\lambda} (v(\lambda) - \mu_v)(v'(\lambda) - \mu_{v'})}{\sigma_v \sigma_{v'}} \qquad (12)$$

3.4 Experimental Analysis

Dimensionality reduction was performed on the selected datasets, i.e., Indian Pines, Pavia University, and Salinas, using two techniques PCA and FastICA. For the experimental purpose, Python 3.7.4 (Jupyter notebook) and Google Colab GPU was used. Figure 10 shows the hyperspectral image of all the three datasets.

The explained variance ratio given by a principal component is the balance between the variance of that principal component and the total variance. The explained variance ratio with 2 bands for the datasets are as follows: Indian Pines

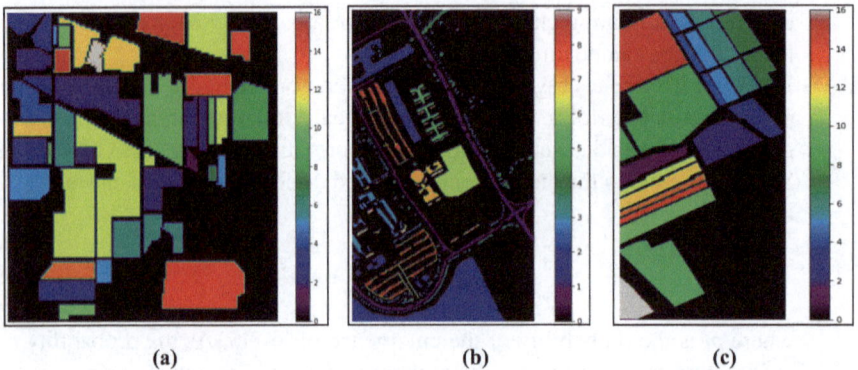

Fig. 10 Hyperspectral image: **a** Indian Pines, **b** Pavia University, **c** Salinas

[0.78141881, 0.21858119]; Pavia University [0.69501038, 0.30498962]; Salinas [0.70640511, 0.29359489]. The ratio also signifies the total distribution of samples over the different principal components. In our case, say for Indian Pines dataset, nearly 78% of data sample is covered by the first principal component and 22% by the second one. The graphs for the explained variance ratio for all the datasets are shown in Fig. 11.

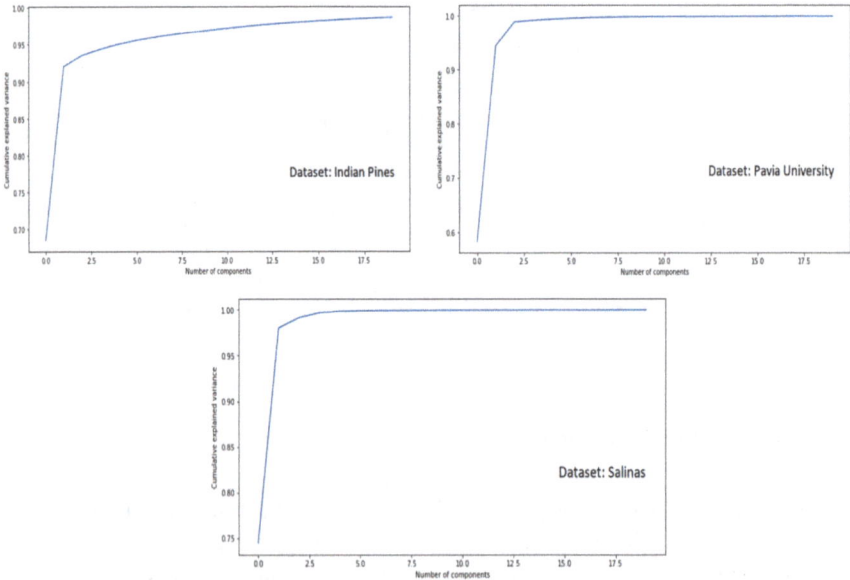

Fig. 11 Explained variance ratio using PCA

The number of spectral channels (or bands) was reduced to eight for each of the dataset. The hyperspectral images of the three datasets for the first eight bands after applying PCA and FastICA techniques are depicted in Fig. 12, Fig. 13 and Fig. 14.

Some well-known dimensionality reduction techniques are also compared below based on the performance and quality criterion mentioned in Sect. 3.3 (as shown in Table 3). Thus, it can be concluded that PCA works well with linear dataset. For nonlinear data, the nonlinear techniques are able to excel the traditional PCA, despite their high variance. Although PCA is widely used in HSI data analysis, it is not a relevant feature extraction technique when the elements of the greatest variation do not correlate with a broad interclass variation [12]. The table below shows that the Sammon technique is quite efficient in most of the criterion as compared to other DR techniques based on artificial hyperspectral data taken.

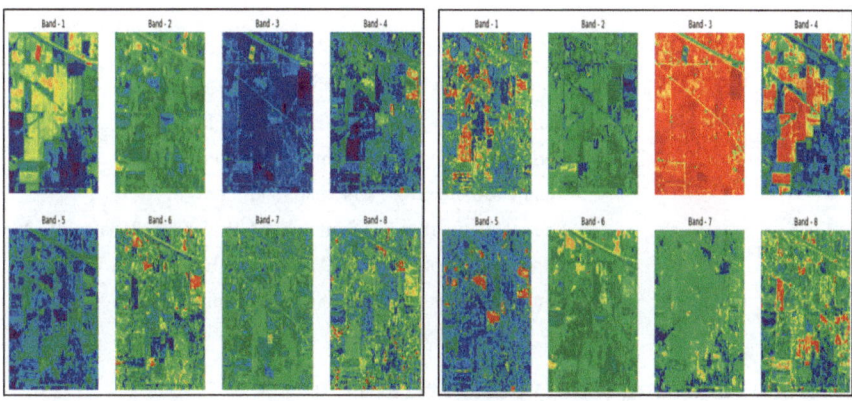

Fig. 12 Indian pines HSI after applying: PCA (left), FastICA (right)

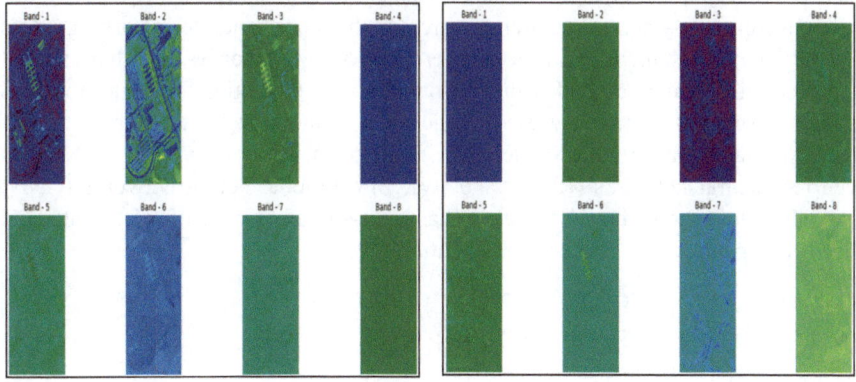

Fig. 13 Pavia university HSI after applying: PCA (left), FastICA (right)

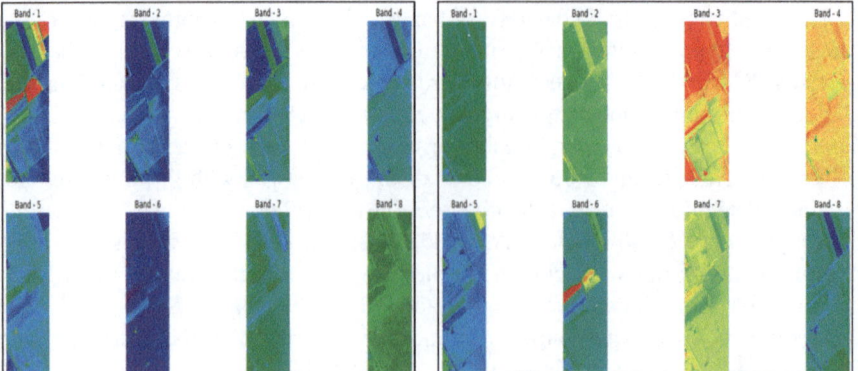

Fig. 14 Salinas HSI after applying: PCA (left), FastICA (right)

Table 3 Relevant DR methods [15]

Quality criteria	Relevant techniques
Entropy	NPE, LLTSA, LLE
Variance	Isomap, Sammon
Mean absolute error	Isomap, Sammon
Spectral derivative	Sammon, MDS, S-SNE, Isomap, PCA
Time calculation	PCA, MDS, LLE, LPP, LTSA,DM
Similarity	PCA, MDS, Sammon

4 Deep Learning for Hyperspectral Data

Conventional machine learning models for classification of hyperspectral data reveal finite restricted performance, using supervised learning of entirely spectral properties within a completely connected architecture. On the other hand, deep learning offers a variety of models, including various layers, exploiting properties in both the spatial and spectral domains, and adopting various learning strategies. Till date, four DL models have been extensively used as the main framework for the exploration of hyperspectral data: (i) autoencoders (AE), (ii) deep belief networks (DBN), (iii) recurrent neural networks (RNN), and (iv) convolutional neural networks (CNN) [19]. Here, it is shown how CNN is used as a classification model for HSI data, and thus is compared with the other classification models.

4.1 Types of Features

The features obtained from the hyperspectral data, $Z \in N^{n_1 \times n_2 \times n_{bands}}$, are identified by their two spatial components $n_1 \times n_2$, and higher spectral domain n_{bands}, enabling the characterization of both types of properties.

(i) **Spectral-based features**: DL models with spectral-based features adopt spectral representations from Z, processing every pixel vector $x_i \in Z$, such that it is fully isolated from the remaining pixels in the image, assuming that each x_i carries a pure and perfect signature of a single surface material, without combining the varying ground cover materials [2]. The performance and predicted accuracy of such classifiers are closely linked to the present training samples and require them to analyze the parameters properly along with the spectral interclass similarity and intraclass variability in order to prevent misclassification of the samples.

(ii) **Spatial-based features**: With progress in remote sensing technology and to overcome the limitation of spectral-based features, the spatial resolution has steadily improved, allowing hyperspectral data cubes to denote target materials with finer spectral pixels and escalating the number of samples collected for each coverage type, and enhancing the acquisition and detection of definite spatial patterns available in ground cover materials, reducing the intraclass variance and uncertainty of the label. Even though these models might outcome the spectral techniques with unique spatial systems and separate spectral signatures in high spatial HIS resolution, the combination of both spectral and spatial properties is further preferable [2]. It includes both the study of spectral signatures as well as the background details associated with it.

4.2 Deep Learning Layers

DL main building blocks are convolutions, downsampling (or pooling) operators, activation functions, and fully connected layers, that are basically identical to the latent layer of a multilayer perceptron (MLP) (Fig. 15).

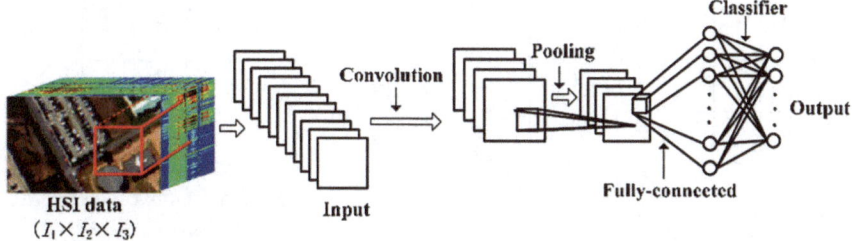

Fig. 15 DL model layers [20]

(i) **Convolution Layer**: A typical layer consists of a group of filters, each to be used on the whole input matrix (or vector), creating a linear amalgamation of all the pixel values in a neighborhood denoted by the filter size. The area that processes the filter is known as a local receptive field, an output pixel value which is an amalgamation of all the pixels generated. In an multilayer perceptron, every neuron produces a single output on the basis of the values generated from the preceding layers, whereas the output value $v(z, i, j)$ is based on a filter z and local data from the preceding layer focused at a location (i, j). The filter sizes preferably used are $5 \times 5 \times t$, $3 \times 3 \times t$ and $1 \times 1 \times t$, where t is the depth of the tensor [1]. Given below is a n input matrix of size $5 \times 5 \times 1$ and a filter of size $3 \times 3 \times 1$, the output convolved feature represents a matrix of size $3 \times 3 \times 1$ with a stride value of 1 (Fig. 16).

(ii) **Pooling Layer**: The pooling layer (or downsampling layer) is often added after a few convolutionary layers that use the maxpooling operator to downsample the image to reduce the vector's spatial dimensionality. This layer has two primary objectives, first it minimizes the data size, and second it is feasible to generate a multi-resolution filter bank by processing the input in varying scale spaces [1]. Other types of pooling operators include the minpooling and the average pooling (Fig. 17). Given below is a $4 \times 4 \times 1$ input matrix, the two $2 \times 2 \times 1$ output matrix represents pooling feature with a stride value of 2.

(iii) **Activation Layer**: Contrary to using a sigmoid function in MLP, the rectified linear function (ReLU) is also used with the function max[0, x] in CNN after convolutional layers or fully connected layers. ReLU eliminates all the negative values, and all positive values are linear. In addition, the parametric ReLU (PReLU) lets less negative features, parametrized by $0 \le \alpha \le 1$ [1]. When the value of parameter α is fixed, i.e., $\alpha = 0.01$, is called a leaky ReLU. Some other common activation functions used include Swish, Mish, Softplus, etc (Fig. 18).

(iv) **Fully Connected Layer**: After using many convolutional layers, the fully connected (FC) layer is included which works in a manner similar to an MLP's hidden layer to adapt weights for classifying the representation. Unlike a

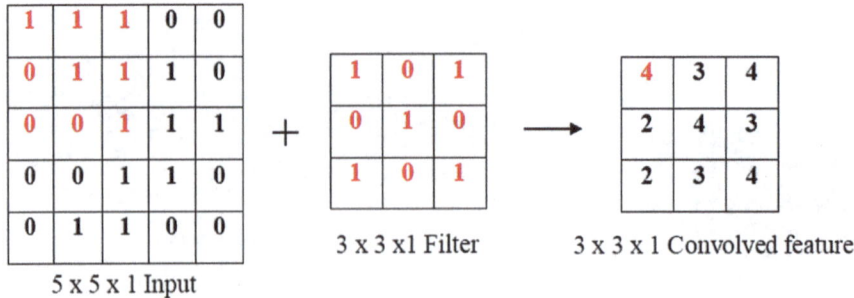

Fig. 16 Convolution process

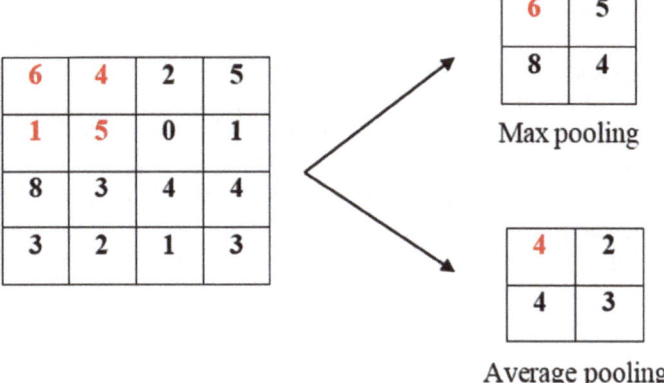

Fig. 17 Pooling feature

Fig. 18 Various activation functions used [2]

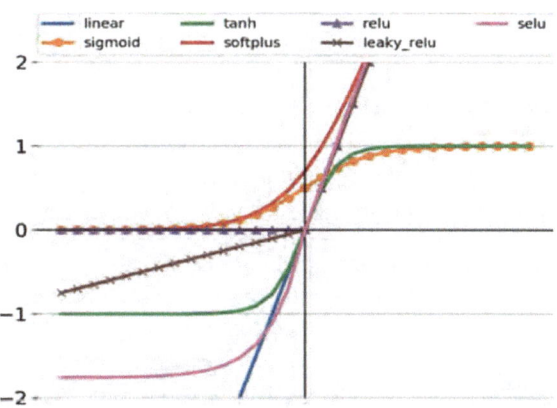

convolutional layer that generates a matrix of local activation values for each filter, the FC layer analyzes the complete input vector, resulting in a single scalar value. For such a purpose, the reshaped version of the data from the final layer is taken as its input. For instance, if the final layer generates a $3 \times 3 \times 30$ tensor, it is reshaped such that it becomes a vector of size $1 \times (3 \times 3 \times 30) = 1 \times 270$. Thus, every neuron in this layer is connected with 270 weights, generating a linear vector combination [1].

4.3 CNN Classification Model

The convolutional layer in the CNN model is the fundamental structural unit, stimulated by the natural process of vision to carry out the extraction of features. In this context, CNN models effectively combine spectral properties with spatial-contextual information from hyperspectral data as opposed to other DNN models in a more effective way. The large pliability it delivers concerning the dimensions of the operational layers, and its aptness to infer powerful supposition regarding the input images have made CNN one of the most accepted and successful classification model, currently being the state of the art in deep learning and an exceptionally desired framework for classification of hyperspectral data. A CNN model is made up of two well-differentiated components which can be explained as two networks: (i) The feature extraction net, which consists of a scaled stack of extraction and identification stages that adopts higher level input representations, and (ii) the classifier that computes the association of every input sample to a particular class made up of a stack of fully connected layers which carries out the last classification job [2]. These coupling networks are together trained as an end-to-end model to enhance all the weights in the CNN. Many CNN models are used for the training and classification purpose, such as 1D, 2D, 3D CNN, and even the hybrid ones. Figure 19 shows a spectral and spectral-spatial CNN model used for classification. Figure 20 shows the step by step classification process of hyperspectral image dataset.

4.4 Performance Evaluation Metrics

In this section, we discuss the various quantitative performance metrics used in the evaluation of a DNN model. The metrics are as follows.

(i) **Confusion Matrix**: A confusion matrix (Fig. 21) is a comparison table generally used for defining the model's output based on a test dataset for which

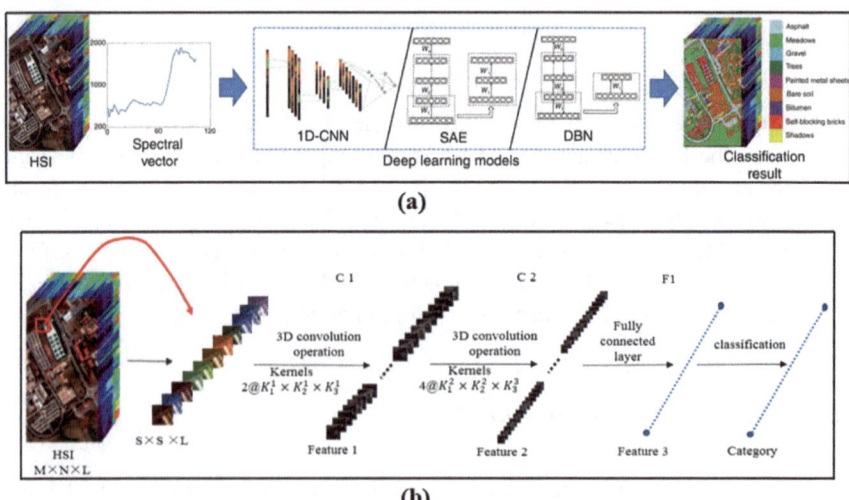

Fig. 19 a Spectral CNN1D architecture, **b** spectral-spatial CNN2D architecture [21]

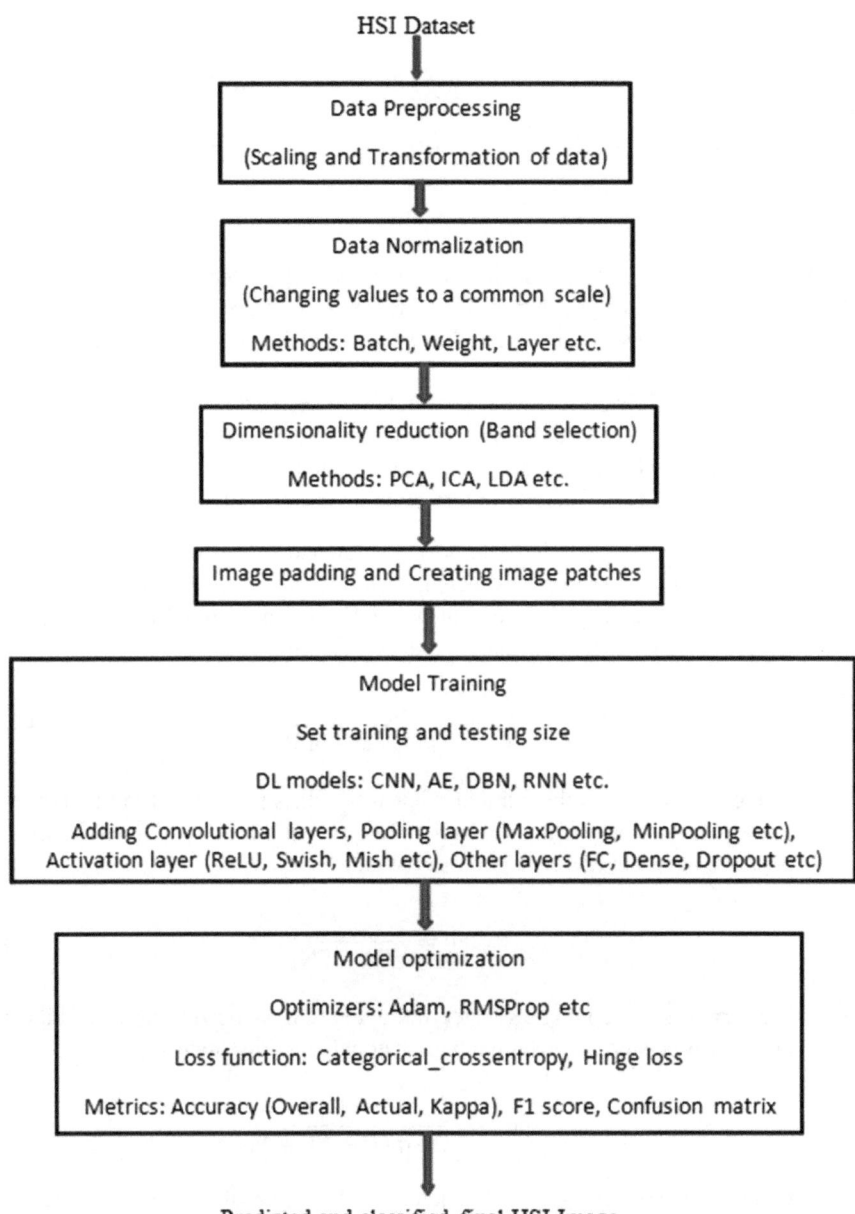

Fig. 20 Step by step classification process

Fig. 21 Confusion matrix

		Actual	
		Positive	Negative
Predicted	Positive	**True Positive**	**False Positive**
	Negative	**False Negative**	**True Negative**

the actual or true values are known. It is basically used to visualize of the performance of an algorithm. Let TP, FP, FN, and TN, respectively, denote the number of true positive, false positive, false negative, and true negative samples [8]. Then,

True Positive (TP): Actually correct, and predicted correct.

False Positive (FP): Actually incorrect, but predicted correct.

False Negative (FN): Actually correct, but predicted incorrect.

True Negative (TN): Actually incorrect, and predicted incorrect [22].

(ii) **Precision**: Precision tells about how accurate/exact the model is out of those positive samples that are expected. Determining precision when the false positive costs are high is a good measure. For example, in HIS classification, a false positive refers to a sample that is misclassified.

$$\text{Precision} = \frac{\text{TP}}{(\text{TP} + \text{FP})} \qquad (13)$$

(iii) **Recall**: Recall gives the number of positive class predictions in the dataset made from all the positive examples. Determining precision when the false negative costs are high is a good measure.

$$\text{Recall} = \frac{\text{TP}}{(\text{TP} + \text{FN})} \qquad (14)$$

(iv) **F-Score**: F-score could be a better metric to use if we need to strike a balance between precision and recall and an unequal class distribution.

$$\text{F-score} = \frac{2\text{TP}}{2textTP + \text{FP} + \text{FN}} \qquad (15)$$

(v) **Overall Accuracy (OA)**: It calculates the ratio of precisely classified hyperspectral pixels by the total number of samples.

(vi) **Average accuracy (AA)**: It calculates the average of the classification accuracies of all the classes.

(vii) **Kappa Coefficient**: It estimates the relation between the acquired classification map and the actual ground truth map. A zero kappa value signifies no agreement, while a one signifies a perfect agreement.

$$K = \frac{N \sum_{i=1}^{n} m_{i,i} - \sum_{i=1}^{n} (G_i C_i)}{N^2 - \sum_{i=1}^{n} (G_i C_i)} \tag{16}$$

4.5 Comparative Analysis

Here, a comparison is made between CNN and the other models used for classification of hyperspectral data. Basically, the accuracies obtained by the spectral (random forest (RF), support vector machine (SVM), multilayer perceptron (MLP), recurrent neural network (RNN), long short-term memory (LSTM), and CNN1D), spatial (CNN2D) and spectral-spatial (CNN3D) techniques are compared. The metrics OA, AA, and kappa are applied on the three datasets discussed in Sect. 2, i.e., Indian Pines, Pavia University, and Salinas Valley, based on their classes, as shown in Tables 4, 5, and 6, respectively. The quantities in bold represent the highest accuracy obtained using that particular classification technique.

From the above data, we can infer that the IP dataset shows better results when CNN2D model is used for classification, whereas the CNN3D model works better for PU and SV datasets giving higher accuracies and kappa coefficient. Strong kappa values for each dataset suggest that the suggested classifier has a higher degree of consonance with the process producing ground truth. Thus, CNN has been a state of

Table 4 IP dataset classification with 15% labeled data [2]

Metrics	RF	SVM	MLP	RNN	LSTM	CNN1D	CNN2D	CNN3D
OA	75.31	84.48	83.50	77.87	83.48	84.02	**99.14**	99.08
AA	61.88	81.05	79.31	70.67	81.18	78.30	98.38	**98.40**
K(×100)	71.41	82.26	81.13	74.65	81.13	81.75	**99.02**	98.95

Table 5 PU dataset classification with 10% labeled data [2]

Metrics	RF	SVM	MLP	RNN	LSTM	CNN1D	CNN2D	CNN3D
OA	89.37	94.10	94.04	92.32	93.0	94.61	98.27	**99.92**
AA	86.02	92.40	92.02	89.95	91.35	92.69	97.11	**99.82**
K(×100)	85.67	92.17	92.09	89.79	90.7	92.84	97.71	**99.89**

Table 6 SV dataset classification with 10% labeled data [2]

Metrics	RF	SVM	MLP	RNN	LSTM	CNN1D	CNN2D	CNN3D
OA	90.12	93.67	93.73	93.59	94.85	95.01	97.94	**99.98**
AA	94.34	96.89	96.87	96.65	97.37	97.51	98.65	**99.98**
K(×100)	88.98	92.94	93.02	92.86	94.27	94.44	97.71	**99.98**

the art in the concept of DNN, providing better and efficient results when compared with other classification models.

4.6 Deep Learning Limitations

The HSI challenges, as discussed in Sect. 2, greatly exacerbate the restriction already disclosed by deep neural network models, that are linked to the intricacy of the classifiers, such as the number of parameters needed for such models. In this section, some of the limitations of DL models are enlisted. DNN training is complex, as optimizing and tuning parameters in those models is a non-convex and NP-complete problem [2], much more difficult to train, without ensuring the convergence of optimization technique. Even, in deeper architectures, increasing the number of parameters frequently leads to several local minima. A higher computational load is expected, which involves costly and memory-intensive approaches, due to the huge number of parameters which must be handled by those models. Supervised models often consume large quantities of training data because of the number of parameters which must be fine-tuned, so they appear to be over fitted in case of less training parameters [23]. In this context, combined with the finite availability of the training samples, the high-dimensional nature of hyperspectral data makes deep neural networks very inefficient in generalizing the spread of hyperspectral data, needing unnecessary changes at the training level, whereas the output on the test data is eventually mediocre.

Further, the desired increase in the precision results is not accomplished by simply piling layers by itself. Indeed, forward propagation suffers from substantial data loss, whereas the backpropagation process faces problems in propagating the activations and gradient signal to all layers with the increase in depth of the network. The gradient vanishes moderately as it traverses every layer of the DNN. In virtualized DNNs, this deterioration becomes very serious and leads to its functional absence. These issues widen the objective function of the model until at each iteration, the model cannot properly change its weights. A drawback is also the black box nature [2] of the training process, which makes the internal dynamics of the model very difficult to interpret. Even though many attempts have been made to imagine the parameters of these models, and to strengthen the extraction of important and illustratable filters, this may impede the design and execution of optimization decisions.

5 Conclusion

Deep learning techniques have transformed image analysis process and have shown to be a robust and efficient aid for analyzing remotely sensed higher dimensional data, modifying their nature to the distinctive features of hyperspectral data. The main attention in this paper was on the concept of hyperspectral data, various dimensionality reduction techniques for reducing their spectral channels and how hybrid deep neural networks are used in effective classification of those hyperspectral images. Given their ability to pull out highly discriminatory features and efficiently exploit the spatial-contextual and spectral data present in hyperspectral cubes, CNN-based models have proved to be highly effective and desirable. In addition, methods such as active learning and transfer learning can also help boost the execution of profound models for training small samples using semi-supervised methods and pre-trained models. Nevertheless, enhancements in computing and hardware technologies also increase the difficulty and scope of the networks in a rapid manner, enabling the fine-tuning tasks achievable within a fair time period [2]. In this context, numerous attempts in hardware accelerated fields have made it feasible to integrate DNN models into embedded processors and GPUs that can balance the load of these networks with the complexity of hyperspectral data effectively.

Acknowledgements The authors would like to thank the Pursue's university MultiSpec site through which the Indian Pines dataset was available, and Prof. Paolo Gamba from the Telecommunications and Remote Sensing Laboratory for providing the Pavia University ROSIS dataset. The authors also gratefully acknowledge the helpful comments and suggestions of the associate editors and reviewers, which have improved the quality of the presentation.

Reference

1. M.A. Ponti, L.S.F. Ribeiro, T.S. Nazare, T. Bui, J. Collomosse, "Everything you wanted to know about deep learning for computer vision but were afraid to ask." 30th SIBGRAPI Conference on Graphics, Patterns and Images Tutorials (2017)
2. M.E. Paoletti, J.M. Haut, J. Plaza, A. Plaza. "Deep learning classifiers for hyperspectral imaging: a review." ISPRS J. Photogrammetry Remote Sens **158** (2019)
3. M.J. Khan, H.S. Khan, A. Yousaf, K. Khurshid, A. Abbas, Modern trends in hyperspectral image analysis: a review. IEEE Access **6**, 14118–14129 (2018)
4. M.S. Kim, S.-I. Tu, K. Chao, "Sensing for agriculture and food quality and safety." Proceedings of SPIE—The International Society for Optical Engineering (2009)
5. X. Li, K. Liu, X. Zhang, "Advanced spatial data models and analyses." Geoinformatics 2008 and Joint Conference on GIS and Built Environment (2009)
6. Y.-T. Chen, "Medical image segmentation using independent component analysis-based kernelized fuzzy c-means clustering." Hindawi, Mathematical Problems in Engineering (2017)
7. N. Audebert, B. Saux, S. Lefevre, Deep learning for classification of hyperspectral data: a comparative review. IEEE Geosci Remote Sens Mag, IEEE **7**(2), 159–173 (2019). https://doi.org/10.1109/MGRS.2019.2912563
8. A. Santara, K. Mani, P. Hatwar, A. Singh, A. Garg, K. Padia, P. Mitra,"BASS net: band-adaptive spectral-spatial feature learning neural network for hyperspectral image classifification." IEEE Trans. Geosci. Remote Sens. **55**(9) (2017)
9. X. Zhang, T. Wang, Y. Yang, "Hyperspectral images classification based on multi-scale residual network." (2020)
10. W. Zhao, S. Du, "Spectral–spatial feature extraction for hyperspectral image classifification: a dimension reduction and deep learning approach." IEEE Trans. Geosci. Remote Sens. **54**(8) (2016)
11. https://towardsdatascience.com/dimensionality-reduction-for-machine-learning-80a46c2ebb7e

12. L.M. Bruce, C.H. Koger, J. Li, "Dimensionality reduction of hyperspectral data using discrete wavelet transform feature extraction." IEEE Trans. Geosci. Remote Sens. **40**(10) (2002)
13. "Pattern recognition and image analysis". Springer Science and Business Media LLC (2015)
14. A. Fejjari, K.S. Ettabaa, O. Korbaa, "Chapter 12 feature extraction techniques for hyperspectral images classification." Springer Science and Business Media LLC (2021)
15. J. Khodr, R. Younes, "Dimensionality reduction on hyperspectral images: acomparative review based on artificial datas." 4th International Congress on Image and Signal Processing (2011)
16. S. Lin, W. Chan, J. Li, Z. Cai, "Liquid chromatography/mass spectrometry for investigating the biochemical effects induced by aristolochic acid in rats: the plasma metabolome." Rapid Communications in Mass Spectrometry (2010)
17. A. Khan, I. Kim, "Sparse independent component analysis with interpolation for blind source separation." 2nd International Conference on Computer, Control and Communication (2009)
18. A. Khana, I. Kim, S.G. Kong, "Dimensionality reduction of hyperspectral images using kernel ICA." Proceedings of SPIE—The International Society for Optical Engineering (2009)
19. R. Hang, Z. Li, Q. Liu, P. Ghamisi, S.S. Bhattacharyya, "Hyperspectral image classification with attention aided CNNs." arXiv:2005.11977v2 [eess.IV] (2020)
20. X. Liu, Q. Sun, Y. Meng, M. Fu, S. Bourennane, Hyperspectral image classification based on parameter-optimized 3D-CNNs combined with transfer learning and virtual samples. Remote Sens. **10**, 1425 (2018). https://doi.org/10.3390/rs10091425
21. Y. Li, H. Zhang, X. Xue, Y. Jiang, Q. Shen, "Deep learning for remote sensing image classification: a survey." WIREs Data Mining and Knowledge Discovery published by Wiley Periodicals, Inc (2018)
22. P.S. Bond, K.S. Wilson, K.D. Cowtan, "Predicting protein model correctness in using machine learning." Acta Crystallogr Sect D Struct Biol (2020)
23. M.E. Paoletti, J.M. Haut, J. Plaza, A. Plaza, "Neural ordinary differential equations for hyperspectral image classification." IEEE Trans. Geosci. Remote Sens. (2020)

Machine Learning and Deep Learning Algorithms in the Diagnosis of Chronic Diseases

Gopi Battineni

Abstract A higher collection of medical documents is a valuable resource to retrieve new and valuable knowledge that can be found through data mining. Deep learning and data mining techniques are user-based approaches to identify hidden and novel data patterns. These highly applicable in identify key patterns among big datasets. At present, these are highly applying in healthcare systems especially of medical diagnosis to predict or classify diseases. Simultaneously, machine learning (ML) can detect and diagnose serious diseases like cancer, dementia, and diabetes. Especially deep learning is one application that highly applicable to the healthcare context is digital diagnosis. Besides, it can detect patterns of individual diseases within patient electronic health records (EHR) and produces feedback on anomalies to the doctor. This chapter presented a brief discussion including ML and deep learning approaches in a clinical context, differentiate between structured and unstructured patient data patterns and provide references to applications of mentioned methods in medicine. Besides, it also highlights performance measures and evaluation used in diagnosis prediction and classification process.

Keywords Data mining · Machine learning · EHR · Medical diagnosis · Pattern identification

1 Introduction

Artificial intelligence (AI) is considered as software engineering which attempts to mimic computers like human behavior. One of the fundamental necessities for any intelligent method is learning. The greatest part of the scientists today concurs that there is no knowledge without learning. Hence, AI techniques like machine learning (ML) and deep learning are important parts of machine intelligence.

G. Battineni (✉)
Medical Informatics Centre, School of Medicinal and Health Products Sciences, University of Camerino, 62032 Camerino, Italy
e-mail: gopi.battineni@unicam.it

© The Author(s), under exclusive license to Springer Nature Singapore Pte Ltd. 2021   141
M. Bandyopadhyay et al. (eds.), *Machine Learning Approaches for Urban Computing*,
Studies in Computational Intelligence 968,
https://doi.org/10.1007/978-981-16-0935-0_7

From the earliest starting point, machine learning models are designed to evaluate the clinical information. Today these techniques are becoming fundamental tools to do insight analysis of medical data [1]. Particularly, since the last decade, the computerized transformation gave relatively less expensive and available to accumulate the medical information. Nowadays, hospitals are largely equipped to collect and monitor data from information technology (IT) systems. Moreover, this information is largely gathered from big data frameworks. Given this, machine learning algorithms are better suited to the examination of medical data, and specifically, there is a great deal of work that has been done in clinical analysis especially for diagnostic issues [2].

Deep learning is a subcategory in machine learning algorithms that are learned by artificial neural networks (ANN). These algorithms are applied in self-driving vehicles and automatic voice assistants like SIRI or ALEXA [3]. Deep learning plays a vital role in voice control in consumer goods such as mobiles, TVs, tablets, and Wi-Fi connected speakers. These algorithms conduct feature selection algorithms to detect image or voice characteristics [3, 4]. The association between AI technologies including machine learning and deep learning is presented in Fig. 1. In clinical practice, deep learning algorithms have the capability of producing high disease prediction accuracy which outperformed than human intelligence. ANN models contain many layers and are trained by large samples of labeled data.

Medical diagnostics are a classification of medical tests, which intends to identify infectious diseases, conditions, and ailments. These clinical diagnostics fall under the class of in-vitro medical diagnostics (IVD) that can be purchased by end-users or utilized in research center settings. Biological samples are disengaged from the human body, for example, blood or tissue to give results. Because of the multiple opportunities for utilization of ML in medical diagnosis, clinical imaging work processes are well on the way to be affected in the near term. ML-driven methods that autonomously process 2D or 3D image scans to recognize clinical signs (like

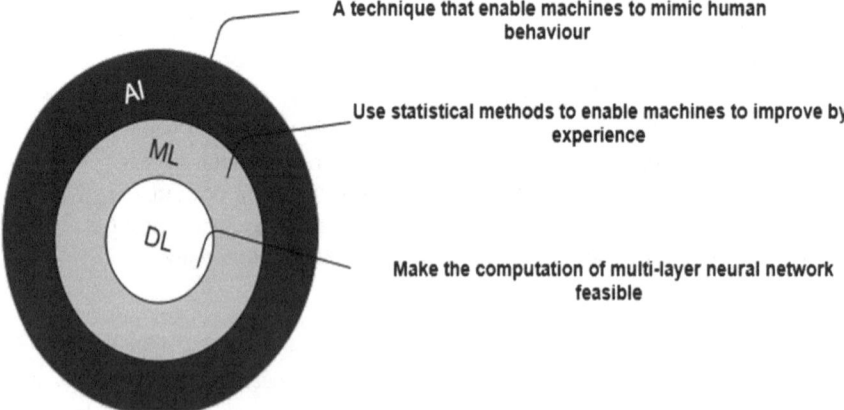

Fig. 1 Relationship between AI, ML, and deep learning

lesions or tumors) or decide possible diagnosis determinations have been distributed, and some are advancing through administrative steps thrives the market.

Among ML, deep learning techniques are largely utilized on layered portrayals of different features, such called neural networks. To understand the deep learning techniques, powerful image data is required to perform recognition tasks [5]. For instance, if a person entered into a dark room and searching for the light switch. From previous experience, the person can figure out how to connect light switches with unsurprising areas inside the design of a room. Numerous computer vision-based picture handling calculations, including deep learning, imitate this behavior to assess factors, which related to the recognition task that needs to be done. Due to the consideration of the multiple complexities of factors, deep learning has its capacity for image interpretation, especially in clinical practice.

The historical progress associated with ML applications in clinical analyze are showing that easy form and straight forward to manage algorithms, frameworks, and approaches have developed to empower progressed and modern data analysis. Both deep learning and machine learning are largely integrated with data mining techniques [6]. Data mining has significance concerning finding the designs, anticipating, and disclosure of information in various spaces. Data mining algorithms and techniques like clustering, classification makes a difference in finding the patterns to choose what has to come to business structures to develop. It is a wide application area nearly in each industry where the information is produced because data mining is considered as one of the most significant frontiers in database and data frameworks, and one of the most encouraging interdisciplinary improvements in IT management. In the next sections, the authors present the importance of ML, and deep learning techniques in the context of medical diagnosis.

2 Machine Learning Framework and Performance Metrics

The simple machine learning framework in medical diagnosis is explained in Fig. 2. The framework including seven individual steps to evaluate the final disease diagnosis and each step was further explained in detail.

Data collection: The model accuracy was decided by the quantity and quality of input medical data. The outcome of the data collection step represents the data used for training purposes. The medical datasets that available from the UCI repository, Kaggle, etc., were collected.

Data preparation: Because of the advancements in the IT industry, high volumes of information are collected from different industries. An IT and database research have offered arise of a way to deal with store and control this important data for decision making in future purposes. Data preprocessing or preparation was identified patterns and key information from these large datasets. This step involves the collection of medical data sets to conduct model training and testing. Data cleansing was involved in duplicate removal, normalization, and error correction.

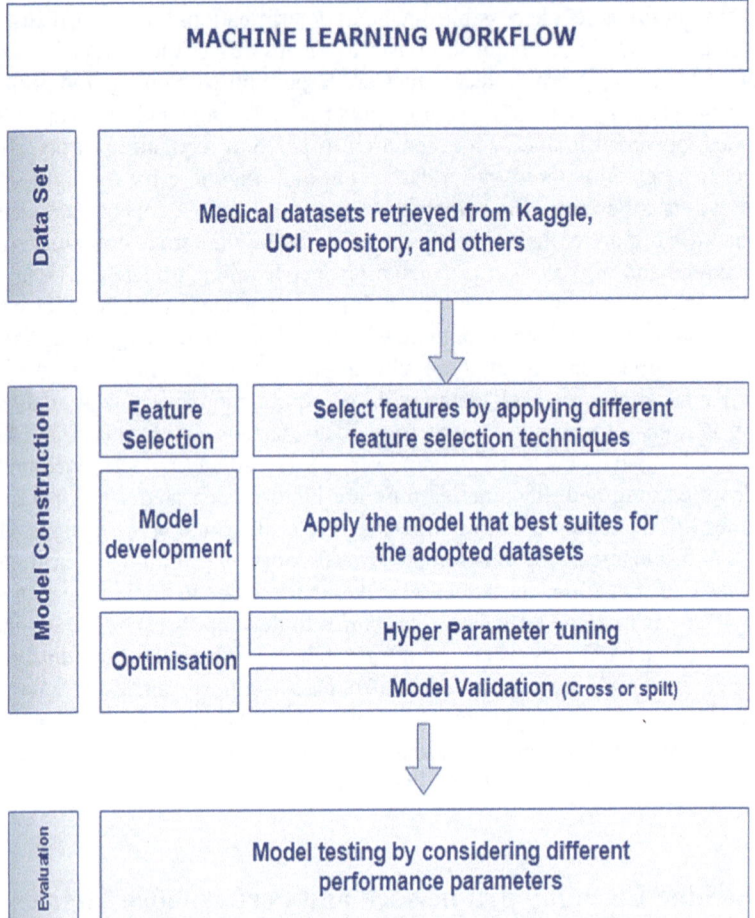

Fig. 2 Machine learning framework in medical diagnosis

Model selection: Machine learning involves several models that are available to do both regression and classification tasks. In this step, we select a singular model among the group of ML models for dataset training. Especially in medical diagnosis, selecting the correct model is important because every model was designed to perform different tasks.

Model training: In this step, the chosen model is properly trained to make disease prediction with the highest accuracy. For example, in cancer diagnosis, the linear regression algorithms are used to retrieve patient type with malignancy or not.

Model evolution: It is necessary to evaluate the machine learning algorithm before adoption into the medical domain. After feature reduction or data preprocessing, and model development, we need to evaluate whether a particular model is accurately identified disease. Different performance metrics are available to assess different

Fig. 3 Confusion matrix

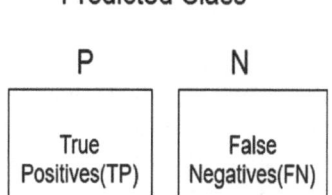

machine learning algorithms. The accuracy and area under the curve (AUC) are used for disease classification purposes, and parameters precision, sensitivity, and F1 scores are used for sorting purposes. Figure 3 presents the confusion matrix example to define the model performance.

From the above confusion matrix, we define the following performance metrics.

Accuracy (A): Portion of actual prediction subjects among total subjects;

$$A(\%) = \frac{\text{True Positives} + \text{True Negatives}}{\text{TP} + \text{TN} + \text{FP} + \text{FN}} * 100$$

Sensitivity (s): True positive subject percentage; $s(\%) = \frac{\text{TP}}{\text{TP}+\text{FN}} * 100$.

Precision (p): Percentage of true positives from total positives; $p(\%) = \frac{\text{TP}}{\text{TP}+\text{FP}} * 100$.

F-Score: Weighted harmonic mean for precision and sensitivity; i.e.,Γ - score $= 2 * \frac{sp}{s+p}$.

The area under curve (AUC): AUC is the visualization tool of multiclass classifier performance and is one of the major evolution parameters to check any classification model performance.

Hyperparameter tuning

Hyperparameter tuning helps to improve model performance. The hyperparameters include distribution and initialization values, training steps, learning rates, etc.

Model predictions

By using test data, the model conducts the classification of medical label data, ultimately validate, and better approximation to verify how the model was performed on real-time medical diagnosis data.

3 ML Algorithms in Clinical Practice

Health issues caused especially by chronic diseases are the main reason for world-wide medical costs. Individuals that suffered by these diseases required permeant treatments. To do this, ML models were frequently applying in the prediction, classification, and diagnosis of different chronic diseases. In this section, the authors discussed some contribution studies involving ML model applications in the primary diagnosis of some chronic diseases.

3.1 Cancers

Due to the capability of detecting hard to discern patterns from complex and noisy datasets, ML algorithms were often used to detect the cancer maladies. Several studies were attempted machine leaning techniques for prognostication and risk prediction of different cancers. Alabi et al. [7] present the uncertainties of relapses in the initial stage of oral tongue squamous cell carcinoma (OTSCC) in decision making of oral lounge cancers. The authors were collected 311 patients' data from five different university hospitals and compared the performance of four ML algorithms, namely naive Bayes (NB), boosted decision tree (BDT), support vector machine (SVM), and decision forest (DF). Preliminary outcomes highlighted that BDT has generated the highest accuracy of 81% and 0.78 value of F-score and SVM generated the lowest accuracy of 68% and 0.63 value of F-score, respectively.

Manabu et al. [8] conducted machine learning algorithms by digital slide images to do the early prediction of colorectal cancer (CRC) metastasis on 397 subjects. A few morphologic boundaries were separated from entire slide pictures of cytokeratin immunohistochemistry images. A random forecast (RF) model was employed by doing data split as a trained dataset of ($n = 277$) images to predict lymph node metastatic also test dataset of ($n = 120$) images. The performance outcomes were further compared to machine learning models and conventional approaches of datasets. Ultimately, lymph node metastatic prediction by ML algorithms was outperformed than other conventional models.

Bikesh Kumar [9] determines breast cancer biomarkers to conduct predictions by anthropometric and clinical features including ML algorithms. Feature correlation and selection methods were employed to evaluate the correlation between different features. Moreover, famous classifiers, for example, SVM, NB, quadratic discriminant, linear discriminant, K-nearest neighbors (KNN), RF, and logistic regression (LR), are introduced for breast cancer predictions. Results highlighted that among the glucose, age, and resisting are seen as generally important and viable biomarkers for malignant growth prediction. Further, the KNN classifier accomplishes the highest 92.1% of classification accuracy.

Raghava et al. [10] did the predictive analysis on total pathological response following Neoadjuvant Chemotherapy to detect breast cancers using ensemble machine learning. Term ensemble learning defines the process of combining multi-models. The results are further validated by K-fold cross-validation and generated 99.08% of accuracy. In similar, Leili et al. [11] did a performance comparison of six ML algorithms to classify survivors of breast cancers and metastasis. Among 550 patients, 85% of them, not experience metastasis, and 83.4% were alive. In a prediction analysis of survival, the SVM produced the highest 93% accuracy. For the prediction of metastasis, the logistic regression generated the highest 86% of total accuracy.

3.2 Alzheimer's

Based on the global Alzheimer's disease (AD) report [12], an individual exposed to dementia will be born every three seconds around the globe. At present, nearly 55 million global population is suffered from this disease. These numbers could be tripled by 2050 and reached 152 million population will be suffered by dementia. AD largely occurs in older people and having a great impact on their daily lives. Early identification of this issue will save medical costs and provides a healthy individual lifestyle. Luckily, because increasing numbers in ML algorithms are well positioned to control this disease at an early phase.

For instance, by collecting speech data from older dementia patients it is possible to identify AD by incorporating ML methods. For this logistic regression with cross-validation, algorithms are the optimal solution to predict early dementia [13]. A comprehensive ML algorithm was developed in [14] by combining four ML algorithms (SVM, NB, KNN, and ANN) to achieve higher accuracy for the early diagnosis of AD. The outcomes of the developed algorithm produced 99.1% accuracy during the identification of true AD subjects. In similar, Battineni et al. [15, 16] successfully presented a study on the prediction of dementia and performance validation by exploring the use of SVM and decision tree algorithms.

All the above-mentioned studies involved model development by medical data retrieved from IT systems and can predict only a single outcome. In contrast, Charles et al. [17] adopted an unsupervised ML approach, namely a conditional restricted Boltzmann machine (CRBM) to evaluate the AD development. The authors presented 18-month projections on 44 clinical characteristics after 1909 AD patients. The presented unsupervised model was accurately predicted the alternations of the AD cognitive scale and identified subcomponents with better sensitivity. Javier et al. [18] presented a sophisticated approach to predicting late-onset AD, and experimental models perform about 72% of classification accuracy.

Besides, Ramon et al. [19] were analyzed the possibility of assessing an anatomical index which is known as AD risk factors. The neuroimaging databases were collected, further applied high-dimensional ML algorithms and results identified cognitive status, age, and cognitive function are the main factors to support the risk of getting AD. In this manner, there are plenty of other studies like [20, 21] were explained the importance of ML models to produce promising results to estimate the genetic risk of AD. However, a survey on ML algorithms for longitudinal analysis of brain image data of AD was well presented in [22].

3.3 Diabetes

Diabetes is a typical continuous medical problem happening when the pancreas has not delivered sufficient insulin. Raised glucose levels are the common consequences of serious diabetes. Thereafter, diabetes will make serious harm to the nerves and veins. Propelled diabetes is problematic by coronary ailment, visual impedance, and kidney disappointment.

As per the World Health Organization (WHO) reports, about 425 million individuals are globally suffered because of diabetes. Number studies are reported that family history, unhealthy diets, hypertension, lack of physical activity, and obesity are risk factors for getting Type 2 diabetes. Women are having a high tendency to get risk for type 2 diabetes because of a high number of pregnancies, low insulin consumption, and high cholesterol levels [23]. Early recognition of the sickness can offer patients the chance to make the fundamental way of life changes, and along these lines can improve their future [24]. To do that computer scientists are recommending cost-effective ML and data mining techniques for the diagnosis of diabetes.

Few investigations lead expectation examination utilizing ML algorithms to analyze diabetes. Nahla et al. [25] involved SVM algorithms to detect diabetes mellitus and achieved 94% accuracy. Moreover, Quan et al. [26] employed J48 decision trees, random forests, and neural networks. Scholars mentioned RF are an ideal algorithm to produce better accuracy (of 80.4%) in the classification of diabetic patients.

Deepti et al. [27] proposed a predictive model to estimate the probability of diabetes. Scholars employed NB, J48 decision trees, and SVM algorithms and concluded that NB generated the highest 76.3% of accuracy than others. On the other hand, Battineni et al. [28] predicted causes for diabetes in Pima Indian female patients by comparative machine learning study. Plasma glucose concentration was the major cause of diabetes happening in these female groups, which is followed by other risk factors like multiple pregnancies and high insulin release. The study by Chaki et al. [29] had provided a systematic investigation of AI and ML approaches for self-management of diabetes mellitus and identification.

3.4 Hepatic Fibrosis

Hepatic fibrosis addresses the injury fixing reaction to the liver from a wide assortment of etiologies. Cirrhosis is the high exceptional phase of fibrosis, implying more than fibrosis alone, yet rather a contortion of the liver parenchyma related with septate and knob arrangement, adjusted bloodstream, and the likely advancement of liver disappointment. Real-time tissue elastography (RTE) is one among contemporary techniques and promising imaging methods since it is both non-obtrusive and gives precise assessments of hepatic fibrosis. It is reported that pattern recognition approaches and machine learning models were largely studied in early diagnosis of hepatitis diseases especially for clinical figures and biochemical indices.

Yang et al. [30] utilized four conventional ML models, namely NB, KNN, SVM, and random forests, to develop the clinical decision system to measure the hepatitis B. Eleven RTE image characteristics are retrieved from 513 patients with liver biopsies. The test results indicated that the employed ML models successfully outperformed the liver fibrosis index (LFI) technique and the random forest (RF) model produced the most noteworthy normal precision among the four ML algorithms. This outcome recommends that modern ML techniques can be incredible assets for assessing the phase of hepatic fibrosis and produce a guarantee for future medical practices.

Jiang et al. [31] were developed a simple model to differentiate patients of clinically significant fibrosis (METAVIR F2-F4) and patients of no or mild fibrosis (METAVIR F0-F1). This study involved 204 community healthcare patients and 34 serum attributes including gender, age, and infection duration that are involved to differentiate fibrosis by the SVM algorithm. Before SVM implementation, feature selection was conducted by a sequential forward floating selection (SFFS) process. Results mentioned that the adopted SVM model to identify patients of clinically substantial fibrosis with 96% accuracy.

Hashem et al. [32], a study has presented single and multistage classification models to predict the degree of liver fibrosis patients with infection caused by chronic hepatitis C. The studies that are previously reported diagnostic techniques not successful to predict early-stage fibrosis irrespective of producing higher accuracies. Given this, scholars of this study developed both single-stage and multistage ML model classifiers to predict the degree of liver fibrosis by employing decision trees, neural networks, nearest neighborhood clusters, and logistic regression models. Preliminary results mentioned that the classification accuracy in means of AUC of multi-model ranging from 0.874 to 0.974 represents improved classifier accuracy than other studies.

Chen et al. [33] presented a sophisticated hybrid model by integrating SVM with local Fisher discriminant analysis (LFDA) for diagnosis of hepatitis. The improved LFDA-SVM algorithm was further compared with the other three conventional methods such as SVM with Fisher discriminant analysis (FDA-SVM), SVM associated principal component analysis (PCA-SVM), and conventional SVM models. The LFDA-SVM model was outperformed than other models and produced 96.77% of accuracy. It is one of the promising and advanced establishing tools in the diagnosis of hepatitis with great performance.

Stoean et al. [34] produced a model by support vector machines to determine liver fibrosis in chronic hepatitis C. The model developed by comprehension of learning components in SVM and the evolutionary algorithms to do engine optimization. By involving evolutionary techniques, it successfully claims better performance than conventional SVM methods and also confirms the significance of new methodology near to reliable support within the medical diagnosis.

Polat et al. [35] did predictive analysis by PCA and artificial immune recognize systems. By reducing the feature set to five from 19 with the help of PCA techniques, the developed system resulted in 94.12% accuracy. Scholars also mentioning that this approach can benefit other medical diagnoses and reduce the doctor's mental stress. By sum up the results of the mentioned studies, machine learning models are the best techniques for staging hepatic fibrosis than other statistical calculations.

3.5 Heart Attacks

Heart attacks are on the head of the deadly ailments list. They are viewed as a major cause of global deaths. As indicated by the WHO statistics, in 2020, about 17 million deaths were caused because of heart diseases [36]. In America, heart diseases such as sudden strokes, hypertension, and coronary heart diseases are the main causes of death. Only because of coronary heart diseases one in seven people lost their lives that resulted in about 366,800 deaths per year in the USA. Approximately, 3% of American adults (7.9 million) are facing the problem of cardiovascular failures. Moreover, a single person dies in every 37 s by heart attacks [37]. Given this, there is a persevering requirement for an exceptionally precise framework that works as an assessment tool to detect hidden patterns of clinical data of heart diseases and reduce the risk of coronary failures. The ML classification algorithms are recently largely incorporated to diagnose heart diseases.

Desai et al. [38] were applied logistic regression and backpropagation neural networks (BPNN) to predict heart diseases for the Cleveland dataset. The models that developed were further validated by tenfold cross-validation. The models are greatly assisting doctors to take effective decisions in the diagnosis of heart failures. Besides, Ahmad et al. [39] proposed a tool to find the better ML algorithm which achieves high accuracy for the prediction of heart diseases. The four classification algorithms such as decision tree, SVM, RF, and logistic regression were employed, and experiments were conducted by total features and selective features. Study outcomes mentioned that the RF classifier algorithm was outperformed than others and produced 94.9% of accuracy.

Enriko et al. [40] developed the heart disease forecasting model by using the K-nearest neighbors (KNN) model with simple patient health features. Authors recommending The KNN with a weighing parameter approach to improve the accuracy in disease diagnosis than others. On the other hand, some authors are proposing hybrid or combined ML algorithms to do a better diagnosis of heart diseases. The hybrid modeling involved two phases. At first, feature selection is conducted and later selected features portion involved in the development of classification models [41]. Similarly, Maji et al. [42] were employed these models by integrating decision trees with artificial neural networks (ANN) to produce better performance in heart disease diagnosis.

Also, Nguyen et al. [43] were applied these algorithms by combining fuzzy models with genetic algorithms for the Cleveland heart disease dataset. The outcomes demonstrated that the genetic algorithms integrated with the fuzzy model are generated better results than other single models like SVM and neural networks. Besides, Manogran et al. [44] also present a system contains neuro-fuzzy and multiple kernel approach inference to do a diagnosis of heart disease. In further, the system was tested by the dataset of metabolic reactions and achieved 98% of sensitivity.

Nazari et.al [45] developed a model by integrating fuzzy inference and the fuzzy analytic hierarchy process (FAHP). Scholars presented a data set of a Tehran city hospital for system training and testing. The fuzzy inference has been used to evaluate the possibility to expose heart problems for individuals and the FAHP for feature weight calculations that contributed to the development of heart attacks.

3.6 Asthma or Chronic Obstructive Pulmonary Diseases (COPD)

Chronic obstructive pulmonary disease (COPD) is a type of lung disease caused by expanding shortness of breath. These diseases are highly involved with morbidity and mortality and is are the third driving reason for death overall in the USA and China [48]. The early diagnosis of COPD patients with future high costs could reduce the medical expenses by exacerbation events and reduce disease evolution. Some studies addressed the ML algorithm to diagnose future high-cost COPD individuals.

For example, Shaochong et al. [49] incorporated smooth Bayesian network (SBN) algorithms to predict the COPD patients involved with future high costs. The developed SBN model aims not only to obtain high prediction accuracy also sophisticated generalizability than other benchmark ML algorithms. In similar, Peter et al. [50] present ML characterization of COPD subtypes by insight analysis from gene study. The longitudinal characterization of COPD gene subjects has provided the relationship between lung image characteristics, COPD progression, and molecular markers.

Moreover, Tadahiro et al. [51] compared the performance measures of different ML algorithms to predict hospitalization and critical care among emergency patients with COPD exacerbation. Outcomes ML algorithms enhanced the prediction ability of patients diagnosed by Asthma or COPD exacerbation. Alternatively, Maarten et al. [52] present a tool to identify functional respiratory imaging features related to COPD and disease forecasting by ML algorithms in early understanding and quantification of disease progression. The FRI features such as total specific image-based airway resistance and volume integrated with SVM algorithms generated 80.65% accuracy and 82.35% of sensitivity.

Sandeep et al. [53] proposed an ML-based framework to evaluate the COPD severity. High correlated features were selected by linear forward feature selection, and KNN was used as a classification algorithm. Results mentioned that the biomechanical feature set outperforms with 0.81 of AUC than density (AUC = 0.71) and texture (AUC = 0.73)-based feature sets. This study provides evidence of the effectiveness of biomechanical features in the severity and the presence of COPD. Likewise, Jianfei et al. [54] also applied a feature weighted survival ML model for the prediction of COPD failures.

3.7 Kidney Injuries

Intense kidney injury is a typical clinical disorder emphatically connected with an abundance of dismalness and mortality. Patients who develop it are at constant risk for delayed and increasingly costly hospitalization, chronic kidney ailment and dialysis, vital adverse cardiovascular situations, and death. Currently, diagnosing kidney failures is made as per the rise of serum creatinine (sCr) focus or decrease in urinary yield, and both are of aberrant markers of renal capacity and may longer days behind the beginning of injury and practical decline. These limitations add to underdiagnosis and identify patients with high risks for kidney injury especially in an emergency condition. Besides, integrating early identification and risk of kidney injury stratification are really engaged support for clinical decisions.

Evaluating the glomerular filtration rate (GFR) is a key parameter to identify initial resistance of kidney functionality, assessing dynamic kidney disintegration and intricacies, altering the measurements of medications, and controlling the risks for chronic kidney diseases. To improve the precision of GFR, the ensemble learning methods were applied [55]. Ensemble learning is a type of ML algorithm that results by combining individual mathematical models to generate better outcomes. Liu et al. [55] employed ensemble learning models and conduct experiments on 1419 individuals. The independent evaluation of GFR was conducted from sex, age, and serum creatinine with the help of SVM, ANN, regression modeling, and ensemble learning. Results mentioning that the precision of ensemble learning dominates the normal regression models.

ML algorithms were also employed to forecast severe kidney injuries after the aortic arch surgeries. For example, Guiyu et al. [56] compared different ML algorithms with conventional logistic regression to predict acute kidney injury (AKI) after arch operations. Results mentioned that the gradient boosting algorithm was comparatively produced better performance than SVM, logistic regression, and random forest algorithms.

Therefore, in this section, we discussed how ML algorithms were applied in the studies of kidney injuries which particularly focused on understanding the association between phenotype and genotype.

3.8 Others

This section presents the discussion on machine learning intervention of other chronic diseases.

The study conducted by Finkelstein et al. [57] defines the ML algorithms to personalize an early diagnosis of asthma exacerbations. Patient telemonitoring brings about an accumulation of huge measures of data about patient illness direction. Authors of this study present comprehensive approaches in the utilization of telemonitoring information for building machine learning algorithms which can forecast the asthma exacerbations before they happen. Experiments were conducted by using adaptive Bayesian networks, Bayesian classifier, and SVM algorithms, and performance in terms of accuracy was achieved 77%, 100%, and 80% respectively. It proves that ML models have the capacity of developing personalized clinical decision support systems.

At the same time, depression is among the main sources of mental illness in developed countries. To adequately target intercessions for patients in danger for a more awful long-haul clinical result, there is a need to distinguish indicators of chronicity and remission at early stages. Because of this, Dinga et al. [58] presented predictive values on a wide range of clinical, biological, and psychological factors in predicting depression causes. They adopted a penalized logistic regression algorithm and archives 66% of prediction accuracy while the diagnosis of course depression.

Besides, muscle pain-related diseases called fibromyalgia (FM) are viewed as a constant, musculoskeletal agony state of clinical unpredictability that presumably emerges from dysfunction for central pain preparing pathways. It is based caused by sleep disturbances, depression or anxiety, and fatigue. But FM diagnosing remains challenging for medical experts because neither laboratory tests nor imaging techniques are not available which can medically confirm or identify the FM diagnosis. To do this, Fred et al. [59] analyse to characterize and differentiate the FM patient's classes from chronic pain patients. In oppose to other established studies that were associated with classification, rather this study included clustering techniques to categorize the pain and symptom severity.

Lastly, periodontitis is an oral disease type driven by deregulated aggravation initiated by polymicrobial networks that structure on subgingival tooth sites. The periodontal pocket and gingival sulcus form unique natural specialties for microbial colonization, and the subgingival microbiota drive the provocative procedure that prompts periodontal tissue destruction. These infection-related diseases were differentiated between chronic and aggressive periodontitis of microbial profiles conducted by support vectors was well explained in [60]. The authors highlight the use of SVM algorithms in the prediction and diagnosis of periodontitis.

In sum up, it is important to present the precise ML algorithms or methods is the most important part to make precise decisions in medical diagnosis [61]. The tabular summary of machine learning studies involved in disease diagnosis has been presented in supplementary material.

4 Deep Learning in Medical Diagnosis

In this chapter, the authors discussed the significance of deep learning in medical diagnosis. Besides, this section presented a brief overview of different scientific publications in the deep neural networks in the medical domain.

The healthcare industry collects large sources of medical information but is not mined to identify the hidden pattern details to make effective decision making. Disclosure of hidden patterns and connections regularly goes unexploited. Comprehensive data mining methods can help to overcome this limitation. Data mining targets a set of given information to identify important and possibly useful patterns. Some example techniques like Bayesian models, artificial neural networks, decision trees, genetic algorithms, and associate rule mining are largely utilized to discover patterns or knowledge, which is previously not known.

Initially, deep learning can use unlabeled data during preprocessing; thereafter, it is well suited for imbalanced datasets and achieves knowledge base [62]. At present these are largely involved in all other problems that not able to address by traditional AI techniques. ANN is the latest deep learning algorithms that are discovered the functionality of different industries. Deep neural networks (DNN) are characterized contributions to profits through a complex composition of layers that presents building blocks including nonlinear functions and transformations. Medical experts feeling that deep learning could be a promising solution in disease identification and symptom detection [63]. There is an expectation which these deep learning, and DNN can alter the chance of getting medical errors like often getting symptomatic errors. Figure 4 presents the deep learning framework of medical diagnosis.

The deep learning techniques can assist the radiologists who specialized in diagnosing the diseases by MRI, computed tomography (CT) scans, and X-rays [64]. There is also the assumption that deep learning can replace human intelligence within the next five years. Because diagnostic imaging holds in medical diagnosis that is usually fit to deep learning models. There are plenty of scenarios that drive the integration of deep learning with number diagnostic practices including radiology; some of them are:

- The ongoing growth of machine intelligence and storage abilities,
- Shortage in healthcare workers,
- Hike in medical costs,
- Large incoming of imaging data, etc.

At present, these algorithms enhance the workflow of the diagnostic process but not mean to replace human intelligence. Below authors explains the promising usage of deep learning in health care in some key medical diseases.

Breast cancer screening

According to the WHO, because of breast cancer, nearly 627,000 global females were facing deaths per year. Medical scientists are recommending mammography for cancer screening at an early phase. In early 2020, Google AI department called deep

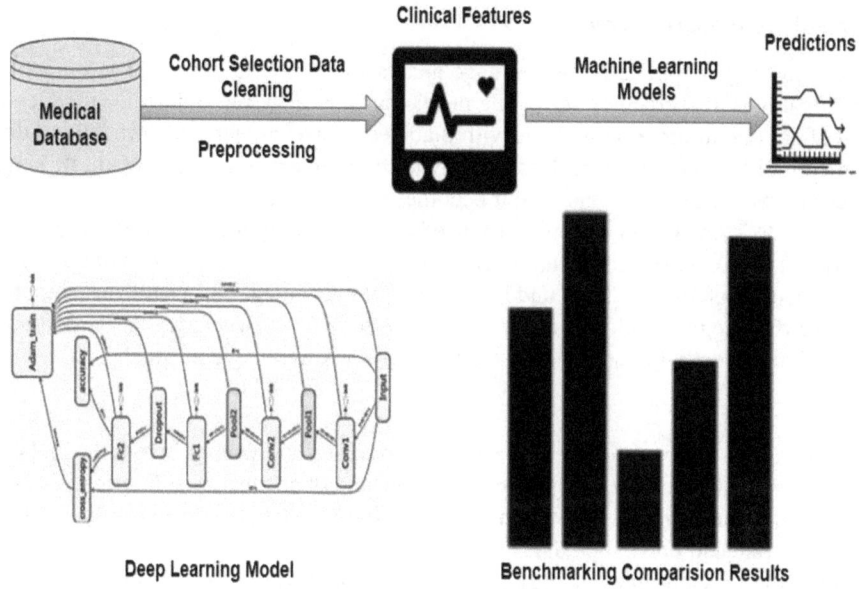

Fig. 4 Deep learning framework of medical diagnosis

mind introduced deep learning models and reported an average 11.5% of percent workload was reduced for radiologists. A research done by the Korean hospitals [65] highlighted neural networks having high sensitivity in cancer detection when compared with human intelligence. Khan et al. [66] developed a novel framework based on deep learning in the detection and classification of breast cancers. The performance of the framework was evaluated by conducting experiments by standard scaled datasets. Results observed that the architecture outperformed than other established deep learning models. However, all these studies are at the beginning stage and improvements had to do with one or more clinical trials.

Melanoma detection

Skin diseases were another significant cause of worldwide disability, and it is the fourth frequent reason that hits skin cancers. These malignancies cured 99% if early detection was done and provide treatment on time. In this time, deep learning can play vital roles because similar to radiologists, the dermatologists also largely depend on visual patterns. In 2017, Codella et al. [67] developed a deep learning system including ensemble machine learning algorithms to analyze tissues in the diagnosis of melanoma. The developed system utilized the dermoscope images including 900 train and 379 testing items. This system enhances the prediction performance in terms of accuracy by 7.5%, precision by 4%, and 2.9 times of sensitivity.

Diabetic retinopathy screening

Deep learning is applicable to do image analysis of retina in particular to diabetic retinopathy (DR) screening [68]. The problems caused by retina lead to blindness and affect one in three individuals with diabetes. However, early identification DR can control the risk of vision loss. However, the problem associated with DR is it does not show any symptoms until it becomes worse to treat.

Given this, Quellec et al. [69] introduced a deep learning model in 2017 to detect early DR and severity classification from mild to moderate with the help of conventional neural networks (CNN), and it achieves 86% accuracy. This result was outperformed by Google developed AI eye doctor with 98.6% of accuracy and proposed algorithm currently helping to Indian doctors from Aravind eye hospital.

Cardiac Arrests

Cardiac arrests are common death causes for both men and women across the globe. An on-time risk assessment conducted with the help of an electrocardiogram (ECG) is the simple and quick examination of heart activities that also significantly reduce the death rate and prevent cardiac arrests.

ECGs are a vital diagnostic tool in the assessment of cardiac arrests. Isin et al. [70] utilized transferred DNN as a feature retrieval method to classify patient ECGs into associate cardiac situations. By doing this, the proposed model has produced an accurate recognition rate of 98.51% and 92% of accuracy. Moreover, Poplin et al. [71] predicted the risk factors for cardiac arrests from retinal images by deep learning models. The models were further trained on 284,335 patients with validation of two individual datasets of 999 and 12,026 patients. The risk factors such as gender (AUC = 0.97), major cardiac events (AUC = 0.70), and smoking habits (AUC = 0.71) significantly helped to predict cardiac diseases. These studies also highlighted the necessity of deep learning models with the training of anatomical features like blood vessels to predict the early cardiac arrests.

Diagnosis of early strokes by head CT scans

The sudden death of brain cells because of low oxygen referred to as immediate stroke is the second highest cause of global deaths. This risky condition needs immediate treatment or diagnosis. Doctors were saying that if there is availability expert assistance within three hours after the rise of the first symptoms can make a faster recovery.

Some studies like Arbabshirani et al. [72] were collected more than 46,000 brain CT scans to develop a model that controls symptoms of intracerebral hemorrhage (ICH). A DNN model was trained by 37,074 scans thereafter 9499 evaluated in unseen studies. Scholars implemented algorithms in daily clinical care and tested for three months and result in a decrement in diagnostic time by 96%. Besides, this algorithm can detect subtle symptoms of ICH which are missed by the radiologists.

According to several studies, deep learning is successfully applied in the diagnosis of ischemic strokes which are caused by LVO (i.e., large vessel occlusion) [73]. These methods can enhance the LVO stroke diagnosis and help to conduct future studies.

5   Conclusions

In this chapter, the authors have presented the state of the art about the importance of machine learning and deep learning techniques in medical diagnosis because of serious chronic diseases and also explain immediate measures to control them. Moreover, different frameworks of ML and deep learning in clinical practice point of view were well discussed. The implementation of these sophisticated learning workflows is much expensive than others includes high computing intelligence. Besides, the financial limitations these AI practices are common in many industries, also healthcare sectors. However, this chapter covers the medical diagnosis of key chronic diseases in a fruitful manner.

Supplementary material: ML studies involved in the prediction of chronic diseases

Category	Author (Ref)	Diagnosis	Models	Outcomes
Cancer	Alabi et al. [7]	The uncertainties of relapses in the initial stage of OTSCC	NB, BDT, SVM, and DF	81%
	Manabu et al. [8]	The early prediction of colorectal cancer (CRC)	RF	Not defined
	Bikesh Kumar S [9]	Biomarkers for breast cancer predictions	SVM, NB, quadratic discriminant, linear discriminant, KNN, LR, and RF	92.1%
	Raghava et al. [10]	Predictive analysis of the total pathological response	Hybrid modeling	99.08%
	Leili et al. [11]	Prediction of breast cancer survival and metastasis	SVM, LR	86–93%
Alzheimer's	Liu et al. [13]	To understand the AD development of patients in an earlier stage	Logistic regression CV	A presented comprehensive method to identify AD
	Battineni et al. [14, 15]	Diagnose the early stages of AD	Hybrid modeling and SVM	99.1% of accuracy
	Charles et al. [17]	Simulate the AD progression	Conditional Restricted Boltzmann Machine (CRBM)	Identified subcomponents with better sensitivity

(continued)

(continued)

Category	Author (Ref)	Diagnosis	Models	Outcomes
	Javier et al. [18]	Predicting late-onset AD	Hybrid modeling	72% of accuracy
	Ramon et al. [19]	Analyzed the AD risk factors	High dimensional ML algorithms	Cognitive status, age, and cognitive function are the main causes behind AD
Diabetes	Nahla et al. [25]	Detection of diabetes mellitus	SVM	94% of accuracy
	Quan et al. [26]	Identification of diabetic patients	J48, RF, and neural systems	80.4% of accuracy
	Deepti et al. [27]	Estimate the probability of diabetes	NB, J48 decision trees, and SVM	76.3% of accuracy
	Battineni et al. [28]	Causes for diabetes in Pima Indian female patients	NB, J48, RF, and LR	Identify the risk factors of diabetes
	Chaki et al. [29]	Self-management of diabetes mellitus and identification	–	Systematic investigation of ML methods in diabetes diagnosis
Hepatic fibrosis	Yang et al. [30]	Improve the hepatitis B stage finding execution	NB, KNN, SVM, and RF	Show guarantee for clinical applications to predict hepatitis B
	Jiang et al. [31]	To distinguish fibrosis	SVM	96% of accuracy
	Hashem et al. [32]	To predict the degree of liver fibrosis	Single-stage and multistage ML model classifiers	87.4–97.4% of classification accuracy
	Chen et al. [33]	Diagnosis of hepatitis	LFDA-SVM, FDA-SVM, PCA-SVM, and SVM	96.77% of accuracy
	Stoean et al. [34]	Determine liver fibrosis in chronic hepatitis C	SVM	Comprehensive methodology near to reliable support within the fibrosis diagnosis
	Polat et al. [35]	To predict the degree of liver fibrosis	PCA Associated ML algorithms	94.12% of accuracy

(continued)

(continued)

Category	Author (Ref)	Diagnosis	Models	Outcomes
Heart strokes	Desai et al. [38]	Predict the heart diseases for Cleveland dataset	LR and BPNN	Effective decisions in the diagnosis of heart failures
	Ahmad et al. [39]	Prediction of heart diseases	Decision tree, SVM, RF, and LR	94.9% of accuracy
	Enriko et al. [40]	Develops the heart disease forecasting	KNN	Improve the accuracy in disease diagnosis
	Maji et al. [42]	Heart disease diagnosis	Hybrid modeling (J48 + ANN)	Combined ML algorithms to do a better diagnosis of heart diseases
	Nguyen et al. [43]	Predict the heart diseases for Cleveland dataset	Combining fuzzy models with genetic algorithms	Hybrid models are generated better results than other single models
	Manogran et al. [44]	Heart disease diagnosis	Neuro-fuzzy and multiple kernel approach	98% of sensitivity
	Nazari et al. [45]	The possibility of getting heart diseases	Fuzzy inference and FAHP	Achieves better accuracy than single classification models
	Jayaraman et al. [46]	Heart stroke classification	Neural networks	Developed the high-performance models by feature selection
	Gokul Nath et al. [47]	Heart disease classification	Support vector machines and genetic approaches	83.7% to 88.34% of accuracy
COPD	Shaochong et al. [49]	Predict the COPD patients involved with future high costs	SBN	Obtain high prediction accuracy also sophisticated generalizability
	Peter et al. [50]	COPD subtypes by insight analysis from gene study	ML with gene algorithms	Relationship between COPD progression, and molecular markers
	Tadahiro et al. [51]	Predict hospitalization and critical care among emergency patients	Lasso regression, RF, boosting, and deep neural network	Enhanced the prediction ability of patients diagnosed with Asthma

(continued)

(continued)

Category	Author (Ref)	Diagnosis	Models	Outcomes
	Maarten et al. [52]	Identify functional respiratory imaging features	SVM	80.65% accuracy and 82.35% of sensitivity
	Sandeep et al. [53]	Asses the COPD severity	KNN	Classification accuracy of 81% was achieved
	Jianfei et al. [54]	Prediction of COPD failures	The new cox ML algorithm	Models proposed develops great promise in clinical applications
Kidney injuries	Liu et al. [55]	GFR evaluation	SVM, ANN, regression modeling, and ensemble learning	Ensemble modeling is a better choice to calculate the GFR with high precision
	Guiyu et al. [56]	Predict acute kidney injury	Gradient boosting, SVM, LR, and RF	Gradient boosting largely increases the prediction accuracy of kidney injuries
Others	Finkelstein et al. [57]	Early diagnosis of asthma	Adaptive Bayesian networks, Bayesian classifier, and SVM algorithms	Bayesian classifier 100% accurately diagnose the asthma
	Dinga et al. [58]	Depression	Penalized logistic regression algorithm	66% of prediction accuracy
	Fred et al. [59]	Muscle pains	Clustering	Characterize and differentiate the FM patient's
	Feres et al. [60]	Periodontitis	SVM	Infection-related diseases were well differentiated

References

1. S. Wang, R.M. Summers, Machine learning and radiology. Med. Image Anal. (2012). https://doi.org/10.1016/j.media.2012.02.005
2. B. Kaur, M. Sharma, M. Mittal, A. Verma, L.M. Goyal, D.J. Hemanth, An improved salient object detection algorithm combining background and foreground connectivity for brain image analysis. Comput. Electr. Eng. (2018). https://doi.org/10.1016/j.compeleceng.2018.08.018

3. A. Voulodimos, N. Doulamis, A. Doulamis, E. Protopapadakis, Deep learning for computer vision: a brief review. Comput. Intell. Neurosci. (2018). https://doi.org/10.1155/2018/7068349
4. Y. Bengio, A. Courville, "Deep learning of representations". Intell. Syst. Ref. Libr. (2013). https://doi.org/10.1007/978-3-642-36657-4_1
5. M. Mittal et al., An efficient edge detection approach to provide better edge connectivity for image analysis. IEEE Access (2019). https://doi.org/10.1109/ACCESS.2019.2902579
6. F. Coenen, Data mining: past, present and future. Knowl. Eng. Rev. (2011). https://doi.org/10.1017/S0269888910000378
7. R.O. Alabi et al., Comparison of supervised machine learning classification techniques in prediction of locoregional recurrences in early oral tongue cancer. Int. J. Med. Inform. (2020). https://doi.org/10.1016/j.ijmedinf.2019.104068
8. M. Takamatsu et al., Prediction of early colorectal cancer metastasis by machine learning using digital slide images. Comput. Methods Programs Biomed. (2019). https://doi.org/10.1016/j.cmpb.2019.06.022
9. B.K. Singh, Determining relevant biomarkers for prediction of breast cancer using anthropometric and clinical features: A comparative investigation in machine learning paradigm. Biocybern. Biomed. Eng. **39**(2), 393–409 (2019). https://doi.org/10.1016/j.bbe.2019.03.001
10. R. Bhardwaj, N. Hooda, Prediction of pathological complete response after neoadjuvant chemotherapy for breast cancer using ensemble machine learning. Inform. Med. Unlocked. **16**(May), 100219 (2019). https://doi.org/10.1016/j.imu.2019.100219
11. L. Tapak, N. Shirmohammadi-Khorram, P. Amini, B. Alafchi, O. Hamidi, J. Poorolajal, "Prediction of survival and metastasis in breast cancer patients using machine learning classifiers," Clin. Epidemiol. Glob. Heal. (September), 1–7 (2018). https://doi.org/10.1016/j.cegh.2018.10.003
12. C. Patterson, "World Alzheimer Report 2018—The state of the art of dementia research: new frontiers," Alzheimer's Dis. Int. London, UK, (2018). https://doi.org/10.1103/PhysRevLett.78.4414
13. L. Liu, S. Zhao, H. Chen, A. Wang, A new machine learning method for identifying alzheimer's disease. Simul. Model. Pract. Theory (2020). https://doi.org/10.1016/j.simpat.2019.102023
14. G. Battineni, N. Chintalapudi, F. Amenta, E. Traini, A comprehensive machine-learning model applied to magnetic resonance imaging (MRI) to predict alzheimer's disease (AD) in older subjects. J. Clin. Med. **9**(7), 2146 (2020). https://doi.org/10.3390/jcm9072146
15. G. Battineni, N. Chintalapudi, F. Amenta, Machine learning in medicine: performance calculation of dementia prediction by support vector machines (SVM). Inform. Med. Unlocked (2019). https://doi.org/10.1016/j.imu.2019.100200
16. B. Gopi, C. Nalini, A. Francesco, Late-life alzheimer's disease (AD) detection using pruned decision trees. Int. J. Brain Disord. Treat. (2020). https://doi.org/10.23937/2469-5866/1410033
17. C.K. Fisher et al., Machine learning for comprehensive forecasting of alzheimer's disease progression. Sci. Rep. (2019). https://doi.org/10.1038/s41598-019-49656-2
18. J. De Velasco Oriol, E.E. Vallejo, K. Estrada, J.G. Taméz Peña, Disease Neuroimaging Initiative TA. Benchmarking machine learning models for late-onset alzheimer's disease prediction from genomic data. BMC Bioinform. **20**(1), 709 (2019). https://doi.org/10.1186/s12859-019-3158-x
19. R. Casanova et al., Using high-dimensional machine learning methods to estimate an anatomical risk factor for alzheimer's disease across imaging databases. Neuroimage (2018). https://doi.org/10.1016/j.neuroimage.2018.08.040
20. E. Moradi, A. Pepe, C. Gaser, H. Huttunen, J. Tohka, Machine learning framework for early MRI-based alzheimer's conversion prediction in MCI subjects. Neuroimage (2015). https://doi.org/10.1016/j.neuroimage.2014.10.002
21. S. Lahmiri, A. Shmuel, Performance of machine learning methods applied to structural MRI and ADAS cognitive scores in diagnosing alzheimer's disease biomed. Sig. Process. Control (2019). https://doi.org/10.1016/j.bspc.2018.08.009
22. G. Martí-Juan, G. Sanroma-Guell, G. Piella, A survey on machine and statistical learning for longitudinal analysis of neuroimaging data in alzheimer's disease. Comput. Methods Programs Biomed. (2020). https://doi.org/10.1016/j.cmpb.2020.105348

23. N.G. Forouhi, A. Misra, V. Mohan, R. Taylor, W. Yancy, Dietary and nutritional approaches for prevention and management of type 2 diabetes. BMJ (2018). https://doi.org/10.1136/bmj.k2234
24. P. Sajda, Machine learning for detection and diagnosis of disease. Annu. Rev. Biomed. Eng. (2006). https://doi.org/10.1146/annurev.bioeng.8.061505.095802
25. N. Barakat, A.P. Bradley, M.N.H. Barakat, Intelligible support vector machines for diagnosis of diabetes mellitus. IEEE Trans. Inf. Technol. Biomed. (2010). https://doi.org/10.1109/TITB.2009.2039485
26. Q. Zou, K. Qu, Y. Luo, D. Yin. Y. Ju, H. Tang, Predicting diabetes mellitus with machine learning techniques. Front. Genet. (2018). https://doi.org/10.3389/fgene.2018.00515
27. D. Sisodia, D.S. Sisodia, "Prediction of diabetes using classification algorithms." (2018). https://doi.org/10.1016/j.procs.2018.05.122
28. G. Battineni, G.G. Sagaro, C. Nalini, F. Amenta, S.K. Tayebati, Comparative machine-learning approach: a follow-up study on type 2 diabetes predictions by cross-validation methods. Machines **7**(4), 1–11 (2019). https://doi.org/10.3390/machines7040074
29. J. Chaki, S. Thillai Ganesh, S.K. Cidham, S. Ananda Theertan, "Machine learning and artificial intelligence based diabetes mellitus detection and self-management: a systematic review." J. King Saud Univ.—Comput. Inform. Sci. King Saud bin Abdulaziz Univ. (4 Jul. 2020). https://doi.org/10.1016/j.jksuci.2020.06.013
30. Y. Chen et al., Machine-learning-based classification of real-time tissue elastography for hepatic fibrosis in patients with chronic hepatitis B. Comput. Biol. Med. (2017). https://doi.org/10.1016/j.compbiomed.2017.07.012
31. Z. Jiang et al., Support vector machine-based feature selection for classification of liver fibrosis grade in chronic hepatitis C. J. Med. Syst. (2006). https://doi.org/10.1007/s10916-006-9023-2
32. A.M. Hashem, M.E.M. Rasmy, K.M. Wahba, O.G. Shaker, Single stage and multistage classification models for the prediction of liver fibrosis degree in patients with chronic hepatitis C infection. Comput. Methods Programs Biomed. (2012). https://doi.org/10.1016/j.cmpb.2011.10.005
33. H.L. Chen, D.Y. Liu, B. Yang, J. Liu, G. Wang, A new hybrid method based on local fisher discriminant analysis and support vector machines for hepatitis disease diagnosis. Expert Syst. Appl. (2011). https://doi.org/10.1016/j.eswa.2011.03.066
34. R. Stoean, C. Stoean, M. Lupsor, H. Stefanescu, R. Badea, Evolutionary-driven support vector machines for determining the degree of liver fibrosis in chronic hepatitis C. Artif. Intell. Med. (2011). https://doi.org/10.1016/j.artmed.2010.06.002
35. K. Polat, S. Güneş, Prediction of hepatitis disease based on principal component analysis and artificial immune recognition system. Appl. Math. Comput. (2007). https://doi.org/10.1016/j.amc.2006.12.010
36. "WHO I The Atlas of Heart Disease and Stroke." https://www.who.int/cardiovascular_diseases/resources/atlas/en/. (Accessed 27 Jul. 2020)
37. S.S. Virani et al., "Heart disease and stroke statistics—2020 update: a report from the American heart association." Circulation. Lippincott Williams Wilkins. E139–E596 (2020). https://doi.org/10.1161/CIR.0000000000000757
38. S.D. Desai, S. Giraddi, P. Narayankar, N.R. Pudakalakatti, S. Sulegaon, "Back-propagation neural network versus logistic regression in heart disease classification." (2019). https://doi.org/10.1007/978-981-13-0680-8_13.
39. H. Ahmed, E.M.G. Younis, A. Hendawi, A.A. Ali, Heart disease identification from patients' social posts, machine learning solution on Spark. Futur. Gener. Comput. Syst. (2019). https://doi.org/10.1016/j.future.2019.09.056
40. I.K.A. Enriko, M. Suryanegara, D. Gunawan, "Heart disease prediction system using k-nearest neighbor algorithm with simplified patient's health parameters." J. Telecommun. Electron. Comput. Eng. (2016)
41. A.L. Chau, X. Li, W. Yu, Support vector machine classification for large datasets using decision tree and fisher linear discriminant. Futur. Gener. Comput. Syst. (2014). https://doi.org/10.1016/j.future.2013.06.021

42. S. Maji, S. Arora, "Decision tree algorithms for prediction of heart disease," in *Lecture Notes in Networks and Systems*, (2019)
43. T. Nguyen, A. Khosravi, D. Creighton, S. Nahavandi, Classification of healthcare data using genetic fuzzy logic system and wavelets. Expert Syst. Appl. (2015). https://doi.org/10.1016/j.eswa.2014.10.027
44. G. Manogaran, R. Varatharajan, M.K. Priyan, Hybrid recommendation system for heart disease diagnosis based on multiple kernel learning with adaptive neuro-fuzzy inference system. Multimed. Tools Appl. (2018). https://doi.org/10.1007/s11042-017-5515-y
45. S. Nazari, M. Fallah, H. Kazemipoor, A. Salehipour, A fuzzy inference-fuzzy analytic hierarchy process-based clinical decision support system for diagnosis of heart diseases. Expert Syst. Appl. (2018). https://doi.org/10.1016/j.eswa.2017.11.001
46. V. Jayaraman, H.P. Sultana, Artificial gravitational cuckoo search algorithm along with particle bee optimized associative memory neural network for feature selection in heart disease classification. J. Ambient Intell. Humaniz. Comput. (2019). https://doi.org/10.1007/s12652-019-01193-6
47. C.B. Gokulnath, S.P. Shantharajah, An optimized feature selection based on genetic approach and support vector machine for heart disease. Cluster Comput. (2019). https://doi.org/10.1007/s10586-018-2416-4
48. C. Wang et al., Prevalence and risk factors of chronic obstructive pulmonary disease in China (the China Pulmonary Health [CPH] study): a national cross-sectional study. Lancet (2018). https://doi.org/10.1016/S0140-6736(18)30841-9
49. S. Lin, Q. Zhang, F. Chen, L. Luo, L. Chen, W. Zhang, Smooth bayesian network model for the prediction of future high-cost patients with COPD. Int. J. Med. Inform. (2019). https://doi.org/10.1016/j.ijmedinf.2019.03.017
50. P.J. Castaldi et al., Machine learning characterization of COPD subtypes: insights from the COPDGene study. Chest (2020). https://doi.org/10.1016/j.chest.2019.11.039
51. T. Goto, C.A. Camargo, M.K. Faridi, B.J. Yun, K. Hasegawa, Machine learning approaches for predicting disposition of asthma and COPD exacerbations in the ED. Am. J. Emerg. Med. (2018). https://doi.org/10.1016/j.ajem.2018.06.062
52. M. Lanclus et al., Machine learning algorithms utilizing functional respiratory imaging may predict COPD exacerbations acad. Radiology (2019). https://doi.org/10.1016/j.acra.2018.10.022
53. S. Bodduluri, J.D. Newell, E.A. Hoffman, J.M. Reinhardt, Registration-based lung mechanical analysis of chronic obstructive pulmonary disease (COPD) using a supervised machine learning framework. Acad. Radiol. (2013). https://doi.org/10.1016/j.acra.2013.01.019
54. J. Zhang, S. Wang, J. Courteau, L. Chen, G. Guo, A. Vanasse, Feature-weighted survival learning machine for COPD failure prediction. Artif. Intell. Med. (2019). https://doi.org/10.1016/j.artmed.2019.01.003
55. X. Liu et al., Improving precision of glomerular filtration rate estimating model by ensemble learning. J. Transl. Med. **15**(1), 1–5 (2017). https://doi.org/10.1186/s12967-017-1337-y
56. G. Lei, G. Wang, C. Zhang, Y. Chen, X. Yang, Using machine learning to predict acute kidney injury after aortic arch surgery. J. Cardiothorac. Vasc. Anesth. (2020). https://doi.org/10.1053/j.jvca.2020.06.007
57. J. Finkelstein, I. Cheol Jeong, Machine learning approaches to personalize early prediction of asthma exacerbations. Ann. N. Y. Acad. Sci. (2017) https://doi.org/10.1111/nyas.13218
58. R. Dinga et al., Predicting the naturalistic course of depression from a wide range of clinical, psychological, and biological data: a machine learning approach. Transl. Psychiatry **8**(1), 241 (2018). https://doi.org/10.1038/s41398-018-0289-1
59. F. Davis, M. Gostine, B. Roberts, R. Risko, J. Cappelleri, A. Sadosky, Characterizing classes of fibromyalgia within the continuum of central sensitization syndrome. J. Pain Res. **11**, 2551–2560 (2018). https://doi.org/10.2147/JPR.S147199
60. M. Feres, Y. Louzoun, S. Haber, M. Faveri, L.C. Figueiredo, L. Levin, Support vector machine-based differentiation between aggressive and chronic periodontitis using microbial profiles. Int. Dent. J. **68**(1), 39–46 (2018). https://doi.org/10.1111/idj.12326

61. G. Battineni, G.G. Sagaro, N. Chinatalapudi, F. Amenta, Applications of machine learning predictive models in the chronic disease diagnosis. J. Personalized Med. (2020). https://doi.org/10.3390/jpm10020021
62. M. Mittal, L.M. Goyal, S. Kaur, I. Kaur, A. Verma, D. Jude Hemanth, Deep learning based enhanced tumor segmentation approach for MR brain images. Appl. Soft Comput. J. (2019). https://doi.org/10.1016/j.asoc.2019.02.036
63. M.A. Khan et al., Gastrointestinal diseases segmentation and classification based on duo-deep architectures. Pattern Recognit. Lett. (2020). https://doi.org/10.1016/j.patrec.2019.12.024
64. A. Mittal et al., Detecting pneumonia using convolutions and dynamic capsule routing for chest X-ray images. Sensors (Switzerland) (2020). https://doi.org/10.3390/s20041068
65. "AI helps radiologists improve accuracy in breast cancer detection with lesser recalls. Healthcare IT News." https://www.healthcareitnews.com/news/asia-pacific/ai-helps-radiologists-improve-accuracy-breast-cancer-detection-lesser-recalls. (Accessed 29 Jul 2020)
66. S.U. Khan, N. Islam, Z. Jan, I. Ud Din, J.J.P.C. Rodrigues, "A novel deep learning based framework for the detection and classification of breast cancer using transfer learning." Pattern Recognit. Lett. (2019). https://doi.org/10.1016/j.patrec.2019.03.022
67. N.C.F. Codella et al., Deep learning ensembles for melanoma recognition in dermoscopy images. IBM J. Res. Dev. (2017). https://doi.org/10.1147/JRD.2017.2708299
68. D.J. Hemanth, J. Anitha, L.H. Son, M. Mittal, Diabetic retinopathy diagnosis from retinal images using modified hopfield neural network. J. Med. Syst. (2018). https://doi.org/10.1007/s10916-018-1111-6
69. G. Quellec, K. Charrière, Y. Boudi, B. Cochener, M. Lamard, Deep image mining for diabetic retinopathy screening. Med. Image Anal. (2017). https://doi.org/10.1016/j.media.2017.04.012
70. A. Isin, S. Ozdalili, Cardiac arrhythmia detection using deep learning. (2017). https://doi.org/10.1016/j.procs.2017.11.238
71. R. Poplin et al., Prediction of cardiovascular risk factors from retinal fundus photographs via deep learning. Nat. Biomed. Eng. (2018). https://doi.org/10.1038/s41551-018-0195-0
72. M.R. Arbabshirani et al., Advanced machine learning in action: identification of intracranial hemorrhage on computed tomography scans of the head with clinical workflow integration. npj Digit. Med. (2018). https://doi.org/10.1038/s41746-017-0015-z
73. N.M. Murray, M. Unberath, G.D. Hager, F.K. Hui, Artificial intelligence to diagnose ischemic stroke and identify large vessel occlusions: a systematic review. J. NeuroInterventional Surg. (2020). https://doi.org/10.1136/neurintsurg-2019-015135

Security Enhancement of Contactless Tachometer-Based Cyber Physical System

Kundan Kumar Rameshwar Saraf, P. Malathi, and Kailash Shaw

Abstract Brushless DC (BLDC) motor truly replaces many brushed DC motors due to its features like high stability, larger torque, less power consumption, and simple control mechanism. Speed and direction of BLDC motor can be controlled by Hall Effect sensor from any remote location using Internet. This Internet of Things (IoT)-based remote control of BLDC motor can be susceptible to many security attacks. These attacks may result in loss of confidentiality, integrity, o r availability of BLDC motor CPS. Hence, it is essential to protect the CPS of BLDC against all these attacks. Initially, this chapter focuses on available BLDC motor speed control systems. Then it demonstrates the CPS of contactless tachometer-based BLDC motor control. To operate this system from remote location, this research has created a webpage. By using this webpage, the user can remotely insert the speed of BLDC motor. Also, the user can know the present speed of BLDC motor by using this webpage. This research discusses possible security threats to this system. To protect the unwanted attacks on this system, the security measures are implemented using lightweight cryptographic algorithms and Splunk Enterprise tool. The security measures are prepared by considering the lightweight nature of the system and cost-effective feature. This method of securing CPS using Splunk alert feature and HTTP Event Collector configuration is novel. The Splunk senses the real time machine data generated by CPS. This data is analyzed to protect the system against undesired Cyber Attacks. Finally, this chapter provides a cost-effective solution of a secure Cyber Physical System of Hall Effect sensor-based BLDC motor. Comparative analysis of many similar existing systems is described in detail. It is concluded that the proposed system is inexpensive, secure, fast and lightweight as compared to existing one.

Keywords BLDC motor · CIA triad · Hall effect sensor · Lightweight cryptography · Machine data · Machine learning

K. K. R. Saraf (✉)
Senior Security Consultant, Capgemini Technology Services India Limited, Talawade, Pune, India

P. Malathi · K. Shaw
D.Y. Patil College of Engineering, Savitribai Phule Pune University, Akurdi, Pune, India
e-mail: viceprincipal@dypcoeakurdi.ac.in

© The Author(s), under exclusive license to Springer Nature Singapore Pte Ltd. 2021
M. Bandyopadhyay et al. (eds.), *Machine Learning Approaches for Urban Computing*,
Studies in Computational Intelligence 968,
https://doi.org/10.1007/978-981-16-0935-0_8

1 Introduction

1.1 Need of System

Due to the absence of commutators and brushes, brushless DC (BLDC) motor has high speed, better performance with high efficiency, less power consumption, better heat dissipation, almost no maintenance, less electromagnetic interference and noise, heavy load and high torque control, quick acceleration and deceleration capability, short duty cycle nature. Due to these huge features of BLDC motor, it i s widely used in aerospace industry [1], for example, NASA uses BLDC motor in Apollo spacecraft's life support blowers. These motors are also used in many industrial processes such as actuation, positioning, manufacturing, and power generation. With the use of BLDC motors, the undersea drones used in marine application become inexpensive and much smaller in size. Sparkless and quieter operations of BLDC motor increase its use in medical applications such as dental drill machines. Due to high peak current handling and faster spinning nature, BLDC motors are also used in robotics such as prosthetic limbs, service robots, and battle bots. To reduce the noise of operation electrical bikes uses BLDC motor into the wheel hub.

For smooth working of all above applications, it is necessary to control the speed of BLDC motor. The speed of BLDC motor can be controlled by changing DC voltage or DC current. DC voltage is directly proportional to the speed of BLDC motor. This DC voltage can be controlled by using power transistor for small power motors. For high power, BLDC motor pulse width modulation (PWM) control method is mainly used. In PWM method, speed of motor can be controlled by using variable voltages. The effective voltage applied to motor is proportional to the PWM duty cycle. The PWM circuit generates appropriate PWM pulses using microcontroller or timer IC. The Hall Effect sensor or infrared sensor or optical encoder senses the actual speed of BLDC motor [2]. The Hall Effect sensor measures the speed of BLDC motor without having any actual contact with the motor shaft. This reduces the friction loss, heat dissipation, and speed reduction problem caused by actual contact-based method of speed of measurement.

Virtual Z-Source multilevel inverter topology [3] can be used to control the speed of BLDC motor. This system faced the unwanted security issues during its operation. Heavy industrial unit is used by this system to control the speed of BLDC motor by resilient directed neural network (RDNN) which results in bulky and complicated nature of system. Closed-loop method of BLDC motor speed control can be used using maximum power point tracker algorithm (MPPT) [4]. Security threats and their countermeasures are explained in this paper. Also, this system uses solar panel to charge its battery. The performance of solar panel is highly degraded during adverse weather condition. Hence, this system becomes ineffective in rainy season. To achieve faultless control of BLDC motor, it is advisable to use redundancy-based approach [5]. But this paper only explains the fault-tolerant control of BLDC motor. Implemented system can be connected to Internet to achieve its control by remote location. Every time this system takes 113 ms to diagnose and identify the fault.

Hence for faultless system, this time can be treated as a wasted waiting period which causes operation lags.

Adaptive network-based fuzzy inference system can be used to improve the speed response of the motor [6]. This system uses dsPIC30F4011-based drive system by which this system becomes bulky and expensive.

The proposed system uses the closed-loop method of BLDC motor speed control. Using BLDC motor connection with microcontroller, Hall Effect sensor and Wi-Fi controller this system comprises the contactless tachometer-based Cyber Physical System. In this system, the PWM pulses are sent to motor by ATmega32A micro-controller. According to the size of pulse width of PWM pulses, the input voltage to BLDC motor changes. This will change the speed of BLDC motor. Hall Effect sensor monitors the speed of BLDC motor. This sensor acts as a feedback to settle a user required field by motor. The resulting speed is received by ATMega32A controller and communicated over a network through Wi-Fi controller. User can observe the present motor speed and easily change the speed of BLDC motor by browsing www. bldcmotorcps.in/remotecontrol/index.html. This system also reads the logs generated by CPS. These logs are observed by Splunk Enterprise Trial version. If any unusual activity is observed in generated log, the Splunk immediately sends an alert to the pre decided user. Also to achieve a better security and confidentiality of messages, this system uses Simon lightweight cryptographic algorithm [7].

The direction of BLDC motor will not affect the operation, and the direction is decided by the system user.

1.2 Problem Statement

The CPS of contactless tachometer-based BLDC motor speed measurement uses Internet to control the speed of BLDC motor. Hence, this system may be suscep-tible to various Cyber Attacks. To protect this system, these Cyber Attacks must be continuously monitored and respective alert for each Cyber Attack should be gener-ated to avoid further damage. This research describes the Cyber Physical System of contactless tachometer. It also explains the method of securing this Cyber Physical System.

1.3 Objective of Research

This research shows the implementation of contactless tachometer-based Cyber Physical System. This system measures the speed of BLDC motor using Hall Effect sensor. This system can be susceptible to various Cyber Attacks. To protect this system against these attacks, the security methods are established and finally the security of system is verified to check the credibility of applied security measures.

1.4 Application of Research

This research is used to measure the speed of BLDC motor or other similar motor using contactless tachometer. This research also helps to protect the similar Cyber Physical System against the unsolicited Cyber Attacks.

Rest of the chapter is arranged as follows: Sect. 2 of this chapter describes the material methods used by the proposed system. Section 3 explains the detail operations of the system. Section 4 shows the result generated by this system along with its comparative analysis with the results of other system. Section 5 provides the detail conclusion of this system.

2   Materials and Methods

2.1 Material

Components for this system are selected on the basis of various criteria such as cost-effectiveness, reliability, and performance. To reduce the bulky structure of this system, it is possible to prepare a single chip on which all components will be mounted. This system uses 8-bit ATmega32A microcontroller. It is high-performance microcontroller with 1MIPS throughput per MHZ. Power consumption and processing speed are optimized with this throughput. This controller is cheaper and sufficient to implement all the required functions along with security. I t has maximum speed of 16 MHz and requires low power for its operation. ESP32-S2 Wi-Fi controller is low-cost reliable product and can be used for small IoT devices. Some of the major reasons of using ESP32-S2 Wi-Fi controller are given in comparison Table 1.

As shown in Table 1, it can be concluded that the ESP32-S2 Wi-Fi controller has moderate price, high security, high clock frequency, and low power consumption features. Hence, use of this Wi-Fi controller is appropriate as compared to other available Wi-Fi controllers.

This system uses Hall Effect sensor A3144. Table 2 shows the comparison of A3144 Hall Effect sensor with other available types [8]. The value given in table is considered when ambient temperature is 25 °C and $V_{cc} = 8$ V as operating voltage. The B_{OP} = Operate point when output turns ON, B_{RP} = Release point when output turns OFF, B_{hys} = Hysteresis value (B_{OP}–B_{RP}).

It can be easily concluded from Table 2, that the Hall Effect sensor A3144 has highest operating point value when output turns ON. It also has highest release point value when output turns OFF. Hence, it is beneficial to use this sensor for the proposed CPS.

The brushless DC motor has many benefits as mentioned in introduction above. The comparison of BLDC motor with other motors is given in Table 3.

Table 1 Comparison of Wi-Fi controllers

Parameter	ESP32-S2	ESP32	ESP8266
Cost in ₹	190/-	752.84/-	129/-
Release Year	2019	2016	2014
Microcontroller	Xtensa single-core 32-bit LX7	Xtensa single/dual-core 32-bit LX6	Xtensa single-core 32-bit L106
Clock Frequency	240 MHz	160/240 MHz	80 MHz
Co-Processor	ULP (RISC-V)	ULP	No
SRAM in KB	320	520	160
ROM in KB	128	448	No
Maximum external SPIRAM in MB	128	16	16
Maximum external flash memory in GB	1	No	No
Time to flight	Yes	No	No
Total No. of GPIO Pins	43	34	16
Touch sensors	14	10	No
ADC	20 (12 Bit)	18 (12 Bit)	1 (10 Bit)
Remote control	Yes	Yes	No
Security	Secure boot flash encryption with 4096-bit OTP	Secure boot flash encryption with 1024-bit OTP	No
Low power consumption	Automatic RF power management 5μA in idle mode and 24uA at 1% duty cycle	10μA deep sleep	20μA

As shown in Table 3, it can be concluded that the BLDC motor has good efficiency, high speed, low maintenance, and low noise capability. The issue with high cost can be overcome by the use of low-cost controller like ATmega32A to control the BLDC motor. All the benefits of BLDC motor are mainly because of the absence of commutator and brushes in it. This research uses BLDC motor with following specifications as shown in Table 4.

2.2 Methods

There are mainly two methods of BLDC motor control. First method is open-loop control and second method is closed-loop control. In open-loop control method, DC voltage is provided to motor in chopping manner. But this method causes problem of current control. In closed-loop control method, the supply voltage is varying

Table 2 Specificity, sensitivity, precision, F1 score, recall, and accuracy values of two classes COVID-19 and normal using VGG16 and transfer learning

Parameter	A3141			A3142			A3143			A3144		
	Min.	Typ.	Max.	Min.	Typ.	Max.	Min.	Typ.	Max.	Min.	Typ.	Max.
B_{OP} at $T_A = 25°C$	50	100	160	130	180	230	220	280	340	70	–	350
B_{OP} at $T_A > 25°C$	30	100	175	115	180	245	205	280	355	35	–	450
B_{RP} at $T_A = 25°C$	10	45	130	75	125	175	165	225	285	50	–	330
B_{RP} at $T_A > 25°C$	10	45	145	60	125	190	150	225	300	25	–	430
B_{hys} at $T_A = 25°C$	20	55	80	30	55	80	30	55	80	20	55	–
B_{hys} at $T_A > 25°C$	20	55	80	30	55	80	30	55	80	20	55	–

Table 3 Comparison of BLDC motor with other similar function motors

Parameter	BLDC Motor	Brushed DC Motor	Induction Motor
Structure	Stator and rotor are made up of permanent field magnet	Stator and rotor are made up of either permanent field magnet or electromagnet	Stator and rotor contain windings with electromagnet
Need of Maintenance	No or very low	Periodic	Low
Characteristic of Speed Vs Torque	Flat	Moderate	Nonlinear
Efficiency	High	Moderate	Low
Speed	High	Moderate	Low
Electrical Noise	Low	High	Low
Need of Controller	Required to control the commutation sequence	Required only for variable speed motors	Required only for variable speed motors
Cost of system	High—Because of controller	Low	Low

Table 4 Specifications of proposed BLDC Motor

Model	GM42BLF 60-130
Rated Voltage	12 V
Rated Speed	3000 rpm
Rated Torque	17.7 Nm
Minimum Torque	0.125 Nm
Peak Torque	53.1 Nm
Rated Current	5 A
Rated Power	38 W
Peak Current	14 A
Rotor Inertia	48 g.cm^2
Length	60 mm
Weight	0.45 kg

dependant on the error signal received by feedback. This method gives more accurate speed control of BLDC motor. The proposed system uses closed-loop control. Hall Effect sensor-based speed control is one of the effective methods for BLDC motor [9–12]. Hence, this research uses this method of speed control.

2.3 Mathematical Modeling of BLDC Motor

Figure 1 shows the stator connection of BLDC motor: BLDC motor has three stator windings. This motor has permanent magnet on its rotor. The voltage equations of BLDC motor are as follows,

$$
\begin{bmatrix} v_a \\ v_b \\ v_c \end{bmatrix} = \begin{bmatrix} R_{st} & 0 & 0 \\ 0 & R_{st} & 0 \\ 0 & 0 & R_{st} \end{bmatrix} \begin{bmatrix} i_a \\ i_b \\ i_c \end{bmatrix} + \begin{bmatrix} L_{\sigma s} & 0 & 0 \\ 0 & L_{\sigma s} & 0 \\ 0 & 0 & L_{\sigma s} \end{bmatrix}
$$

$$
\times \frac{d}{dt} \begin{bmatrix} i_a \\ i_b \\ i_c \end{bmatrix} + \begin{bmatrix} e_a \\ e_b \\ e_c \end{bmatrix} + \begin{bmatrix} U_{N0} \\ U_{N0} \\ U_{N0} \end{bmatrix} \tag{1}
$$

$$
v_a = R_{st} i_a + L_{\sigma s} \frac{di_a}{dt} + e_a \tag{2}
$$

$$
v_b = R_{st} i_b + L_{\sigma s} \frac{di_b}{dt} + e_a \tag{3}
$$

$$
v_c = R_{st} i_c + L_{\sigma s} \frac{di_c}{dt} + e_a \tag{4}
$$

In Fig. 1, three stator phase windings are connected with common node voltage UN0. Basically, UN0 is the neutral point to ground. All three phase windings have same stator resistance Rs and same stator inductance Lσs. The trapezoidal back EMF has created in this motor. This EMF has different value in each of these three windings. Back EMF for phase a is ea, for phase b is eb, for phase c is ec. This motor is rotated with speed wm. The electromagnetic torque generated by motor can be given as,

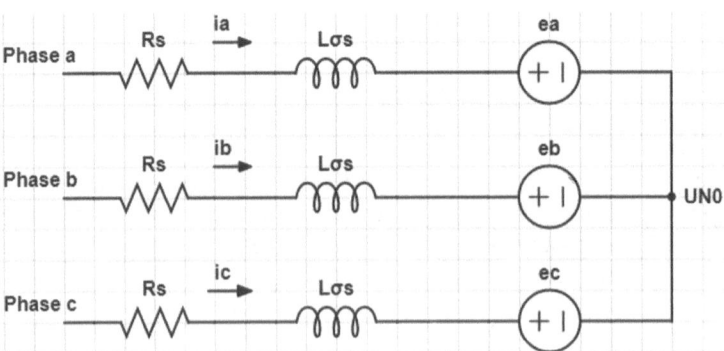

Fig. 1 Stator connection of BLDC motor

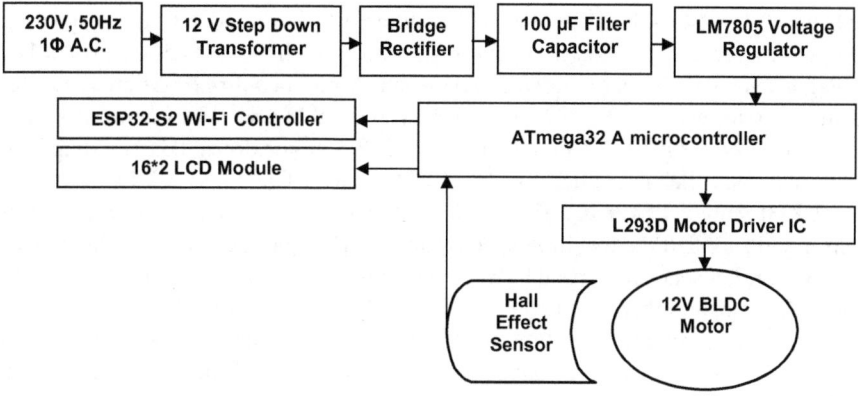

Fig. 2 Block diagram of Contactless Tachometer Cyber Physical System

$$T_e = \frac{e_a i_a + e_b i_b + e_c i_c}{w_m} \tag{5}$$

In BLDC motor, out of its three phases at any time two phases have opposite magnitude of current and third-phase has zero current. This research considers that phase a and phase b has an opposite current and phase c has zero current.

Hence the parameters will be as given below,

$$ia = -ib, ic = 0, ea = -eb \tag{6}$$

The value of electromagnetic torque of Eq. (5) will become,

$$T_e - \frac{2e_a i_a}{w_m} \tag{7}$$

Line voltage between phase a and b is,

$$v_{ab} = v_a - v_b = 2R_s i_a + 2L_{\sigma s}\frac{di_a}{dt} + 2e_a \tag{8}$$

3 Proposed Framework

Figure 2 shows the block diagram of contactless tachometer BLDC motor.

As shown in Fig. 2 Hall Effect sensor measures the speed of BLDC motor [13]. In this figure, single-phase AC supply input 230 V 50 Hz is provided to primary winding of step-down transformer. Secondary of transformer will give 12 V AC voltages. This secondary transformer output is provided to bridge rectifier. Bridge rectifier will give 12 V DC voltages which will be filtered by using 100µF filter capacitor. Output of filter is connected to regulator LM7805 which provides 5 V constant DC voltage. This 5 V DC is provided to ATmega32A microcontroller. Microcontroller is further interfaced with 16*2 LCD display and ESP32-S2 Wi-Fi controller. It is also interfaced with BLDC motor through L293D motor driver IC. The output of this microcontroller is applied to 16*2 LCD module, ESP32-S2 Wi-Fi controller, and L293D motor driver IC. The BLDC motor speed is measured by Hall Effect sensor and supplied the resulting speed value to ATmega32A microcontroller. This microcontroller sends this speed to Wi-Fi controller as well as to LCD module. Wi-Fi controller transfers this speed over a network. User can browse the website www. bldcmotorcps.in/remotecontrol/index.html and observe the present speed of motor as shown in Fig. 4. User can also change the speed of the motor through remote location using the above website [14]. The C ++ language code of this system is given in appendix I: attached with this document. Algorithm and flowchart of proposed system are shown in Algorithm 1 and Fig. 3

Algorithm: Contactless Tachometer Cyber Physical System

Step 1: Start

Step 2: Start the portable hotspot of smartphone named "AccessPoint74"

Step 3: System will be automatically connected to this portable hotspot

Step 4: Open the web page www.bldcmotorcps.in/remotecontrol/index.html on a browser with authenticate id

Step 5: Observe whether the system is connected with the webpage or not?

Step 6: If system is connected then observe the present speed of BLDC motor

Step 7: If system is not connected refresh the webpage and reconnect the system with portable hotspot

Step 8: After the connection of system with webpage established, insert the required speed of the BLDC motor

Step 9: Observe the present speed of system after 0.2 s its speed will be equal to the speed given by user

Step 10: Try to operate the system with any unregistered device

Step 11: Check the alert in the e-mail box which mentions that the "Unauthenticated user tries to access the BLDC motor"

Step 12: Try to perform DoS or DDoS attack

Step 13: Check the alert which mentions that "BLDC motor not generated any logs in last 2 min"

Step 14: Read the corresponding speed on LCD module

Step 15: Stop

As per the program to connect the BLDC motor with external network, any Wi-Fi module or Android hotspot with name "AccessPoint74" is required. In ON condition of this system, it searches for this hotspot. Once it finds the hotspot, it will automatically establish a connection with it. ESP32-S2 Wi-Fi controller is mainly used for this connection of the system with external network. Once the portable hotspot is found, the CPS will establish its connection with the external network. It starts sending its logs. These logs will be detected by Splunk. The Splunk will index the events based on these received logs from CPS. The user needs to first register the IP address of its

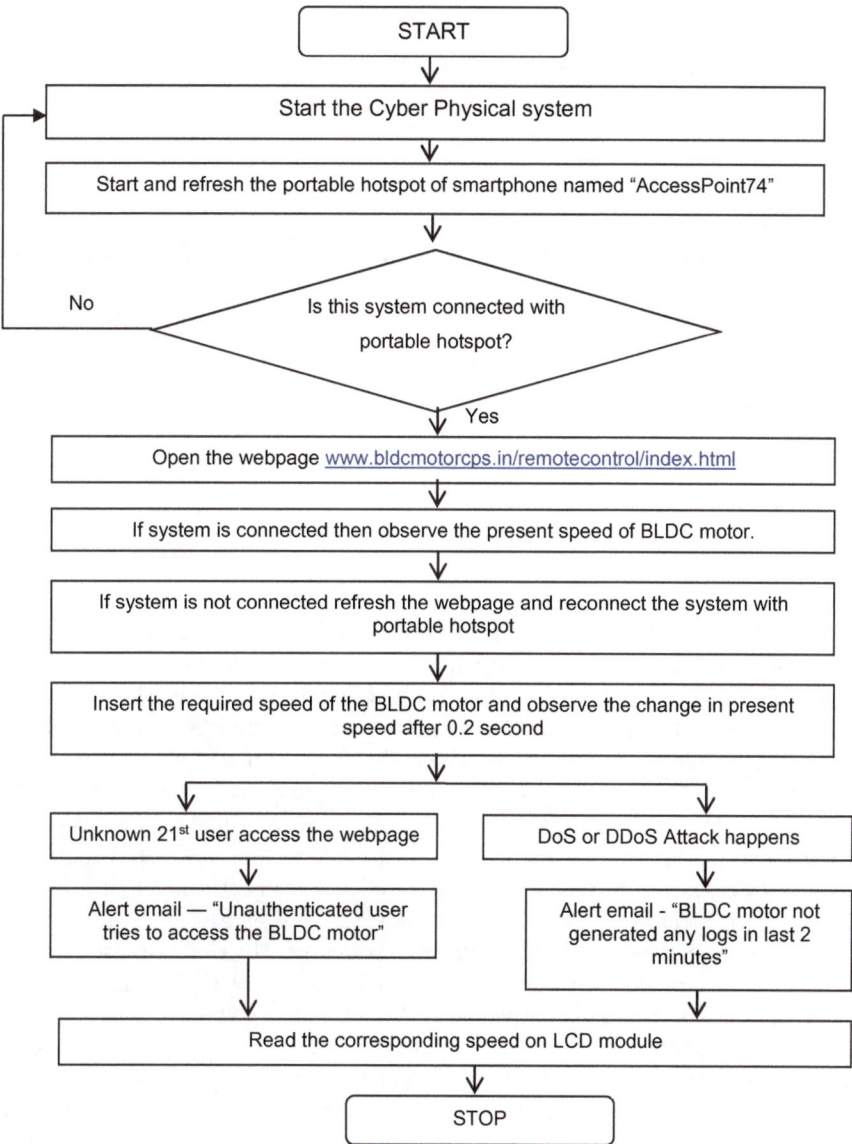

Fig. 3 Flowchart of operation

device into the IP address table of this CPS. User will browse the webpage www.bld cmotorcps.in/remotecontrol/index.html at remote location as shown in Fig. 4. User will change the speed of the motor. Motor will first scan all the stored IP address from its IP address table. If the users IP address matches with anyone of the 20 stored IP addresses, then the motor will accept the request of user and change the speed

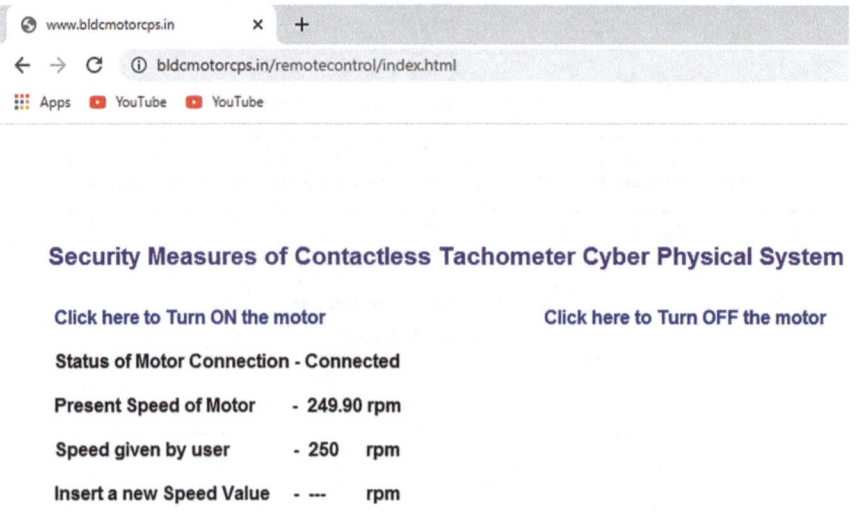

Fig. 4 Webpage to control the system

accordingly. If this IP addresses is not similar to the 20 IP addresses, the Splunk will create an alert and send this alert through e-mail to the owner of this system. The alert created in Splunk is shown in Fig. 8b. Also in this case, the motor control will not be established by this unknown user. If more than five users at a time want to control this system by means of DoS attack the motor will stop its operation. This will be observed by Splunk, and after 2 min, the Splunk will send the alert to the owner. This alert indicates that the BLDC motor is not sending any logs as shown in Fig. 8a. The connected system of BLDC motor, microcontroller, Wi-Fi controller, and Hall Effect sensor is referred as Cyber Physical System. The detail operation of this Cyber Physical System is explained in flowchart 3 (Fig. 3).

In the proposed system, BLDC motor speed is measured by Hall Effect sensor. The ATmega32A microcontroller sends the PWM pulse to motor through L293D motor driver IC. This IC amplifies the voltage. This amplified voltage is sent to BLDC motor to start its rotation. The initial speed of this motor is 100 rpm. One webpage is created to operate this motor from remote location. The address of this webpage is www.bldcmotorcps.in/remotecontrol/index.html. The user can access this webpage and insert any required speed value from remote location as shown in Fig. 4. User can perform the switching operation of motor using this webpage. The motor takes less than 200 ms time to set its new speed as given by user. The Hall Effect sensor measures the present speed of motor, and if it does not equal to the speed given by user, it sends the error signal to ATmega32A microcontroller. This microcontroller tries to achieve a user given speed by adjusting the width of PWM wave. Once the user given speed is achieved by a motor, the error signal received from Hall Effect sensor becomes zero. The ESP32-S2 Wi-Fi controller helps to securely send speed of motor to the user created web page. It also sends the motor speed value given by user to the ATmega32A microcontroller.

4 Software Modification for CPS Security

4.1 Security of Cyber Physical System

The hardware used by this system is explained in Sect. 2.1. This system mainly works on 230 V 50 Hz single-phase supply. This system can be operated on 5 V constant DC power supply. The software used by this system is Splunk Enterprise Version 8.0.3.

To maintain the confidentiality and integrity of this system, all the data over the network is encrypted using Simon lightweight cryptographic encryption algorithm. Also, to maintain the availability, authenticity and authorization of the system, the HTTP event collector (HEC) token is configured by coding this system. Using this HEC token CPS sends its complete log data to the Splunk Enterprise 8.0.3 software. This tool continuously monitors the data received by this CPS. This system is proposed in a manner that only 20 users can operate this system remotely. Before starting the operation of this system public IP address of all these 20 users should be added to the Splunk. The Splunk is configured in such a way that it will create alerts when any unknown IP address tries to operate this system using the above given webpage. The alert message will be sent by e-mail to the authenticate user. On receiving the alert e-mail, the user can take a proper action. Also, this system will not work for the speed given by twenty-first unknown user. By this way, system can perform authorizations and protect itself from this attack. This system should be available every time for its use. Hence, it should be protected against DoS and DDoS attack. Whenever these attacks occur on this system, the operations of system should be immediately stopped. To achieve this goal, the Splunk sends an alert to the authenticate user, if this system stops sending its log data for more than 2 minute of time. Hence, authenticated user will take proper action to avoid these attacks. As this system can only work with 20 users at a time. Hence, it has very less possibility of DoS and DDoS attacks. Also, the 20 public ids are fed to Splunk. Hence, the overhead or overloading of this system is completely avoided by Splunk.

This research is mainly focused toward the security of proposed CPS. This research has implemented a secure CPS of BLDC motor speed control. This system senses the data coming from Hall Effect sensor. Also, user can change the speed of BLDC motor using www.bldcmotorcps.in/remotecontrol/index.html website. This data transfer mainly uses wireless network as its medium of transfer. This system is susceptible to basic attacks such as confidentiality, integrity and availability, authentication and authorization. The details of these attacks and their countermeasures are given below.

Confidentiality: Confidentiality is security mechanism which prevents unauthorised access to data. It also maintains the privacy of information. To increase the confidentiality of data, this system uses lightweight cryptographic algorithm named as Simon encryption algorithm [15].

Integrity: Any information is said to be integrated if it is unmodified during its path from sender to receiver. The data encryption is also useful to maintain the integrity of data [16]. If attacker modifies the encrypted information, the receiver cannot decrypt the message, and hence, data loss is analyzed by receiver. He can send the request of packet retransmission to transmitter in this case. The Simon algorithm also helps to maintain the integrity of data [17].

Simon is lightweight, secure, flexible and analyzable cryptographic algorithm. Simon is mainly used on hardware platforms where confidentiality is a major need.

Simon is designed by classical Feistel scheme. Simon has "2n" bit size of block and "m" size of key. Bijective ness, linearity, and invertible nature are absent in Simon function. This function is applied to each round of Simon. This function is denoted by $F : (F_2)^n \rightarrow (F_2)^n$ to the left half of state. The function F is given by Eq. 9,

$$F(x) = ((x \lll 8) \odot (x \lll 1)) \oplus (x \lll 2) \tag{9}$$

where $x \lll j$ denotes that x is rotated by j number of positions, $\oplus$ denotes bitwise XOR operation and $\odot$ is binary AND operation. The key length m of Simon is multiple of n by factors 2, 3, or 4. The Simon cipher can be obtained by below given formula,

$$\text{Simon Cipher} = \frac{2n}{mn} \tag{10}$$

Simon has various comparative benefits such as it has high security, less memory requirement, small power consumption, and outstanding performance for hardware implementation.

Availability: The availability confirms that the proposed CPS is always accessible to the user. The availability of CPS can be threatened in various cases such as denial-of-service (DoS) or distributed denial-of-service (DDoS) attack on network [18].

Authentication: Authentication is a process of identifying a user based on token or password.

Authorization: It is a permission given by CPS to establish a communication with it [19, 20].

4.2 Steps to Create HEC Token

To overcome the possibility of DoS or DDoS attack, to establish the authenticate communication, and to authorized every user before its connection with proposed CPS, this system uses HTTP Event Collector (HEC) feature of Splunk. Using Splunk Enterprise 8.0.3 version, the HEC is configured [21]. HEC is service provided by Splunk for communication with IoT devices. This service uses token-based authentication for creating HTTP event collector. The steps of use of HEC for proposed CPS are given below.

Only the device which has HEC token and IP address of Splunk Master can perform a communication with Splunk Master. In this research, the logs of the Cyber

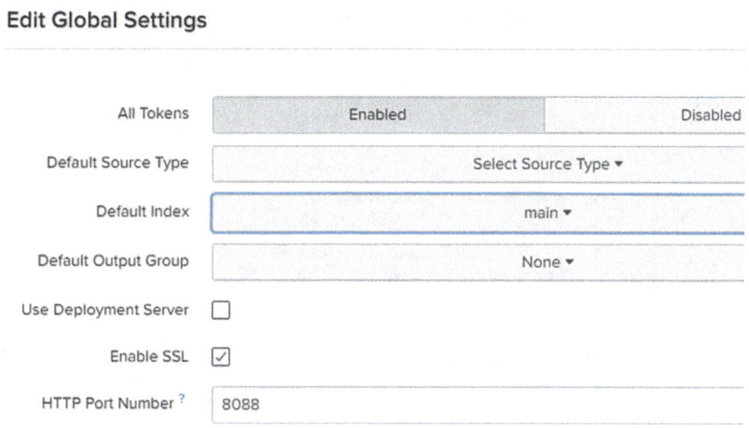

Edit Global Settings

All Tokens	Enabled / Disabled
Default Source Type	Select Source Type ▼
Default Index	main ▼
Default Output Group	None ▼
Use Deployment Server	☐
Enable SSL	☑
HTTP Port Number ?	8088

Fig. 5 Global Settings configuration for before HEC token creation

Physical System are recognized by Splunk Master to detect the unwanted Cyber Attack. The HEC token is only known to the administrator of Splunk Master, and also, it is hard coded in Cyber Physical System. Hence, it is more secure as compared to password authentication.

This token is created following the below steps:

(i) Go to Settings > Data Inputs > Under Local Inputs click on HTTP Event Collector.

(ii) Enable the HEC token by clicking on Global Settings option. Configured the global settings as shown in Fig. 5.

(iii) To create HEC token click on New Token > Fill up the information as given in Fig. 6 > Next > .

(iv) Select main option in the Selected Allowed Indexes > Review > Submit.

This token has a value of **ce8da322-e069-4106-ba76-604d7549519a** as shown in Fig. 7.

4.3 Steps to Establish Connection Between Splunk Master and CPS

The simple one line code is inserted in CPS system to establish its connection with Splunk server. The sample code is given below.

url—k https://IP-Address:8088/services/collector -H 'Authorization: Splunk ce8da322-e069-4106-ba76-604d7549519a' -d '{"event":"This is a test event"}'

To send all Splunk logs, the similar code is used in this research. Any system connected with Internet continuously generates the logs. Hence, various alerts are created in Splunk to monitor the attacks against the security of CPS. The alert is created in Splunk which will be generated, if CPS is unable to send its logs for more

Configure a new token for receiving data over HTTP. Learn More ↗

Name	HEC_Tk1
Source name override ?	optional
Description ?	optional
Output Group (optional)	None ▾
Enable indexer acknowledgement	☐

Fig. 6 New HEC token creation

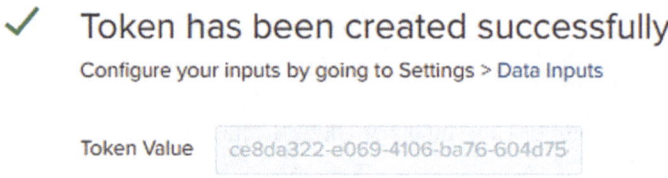

✓ **Token has been created successfully**

Configure your inputs by going to Settings > Data Inputs

Token Value   ce8da322-e069-4106-ba76-604d75

Fig. 7 Token for HTTP Event Collector in Splunk Enterprise 8.0.3

than 2 min. This alert will notify the user that the CPS is unavailable for use. The user will immediately take the required precautionary measures to avoid the threat by this attack against availability. The alert 1 created in Splunk as given in Fig. 8a. To maintain the authenticity and authorization of this CPS, the IP address of user will be prior recognized and registered in Splunk. So Splunk will generate the alert if any new user from unknown or unrecognized IP address tries to use this system. In this case, the owner of this system will recognize the attack on authenticity. The alert 2 creation is shown in Fig. 8b.

5   Result Analysis

Splunk also helps to monitor the performance of CPS. As shown in graph below, the user can change the speed value using android application from remote location. The CPS takes less than 200 ms to analyze the new reading, and it changes the speed of BLDC motor according to new value. Table 5 shows the present reading and user

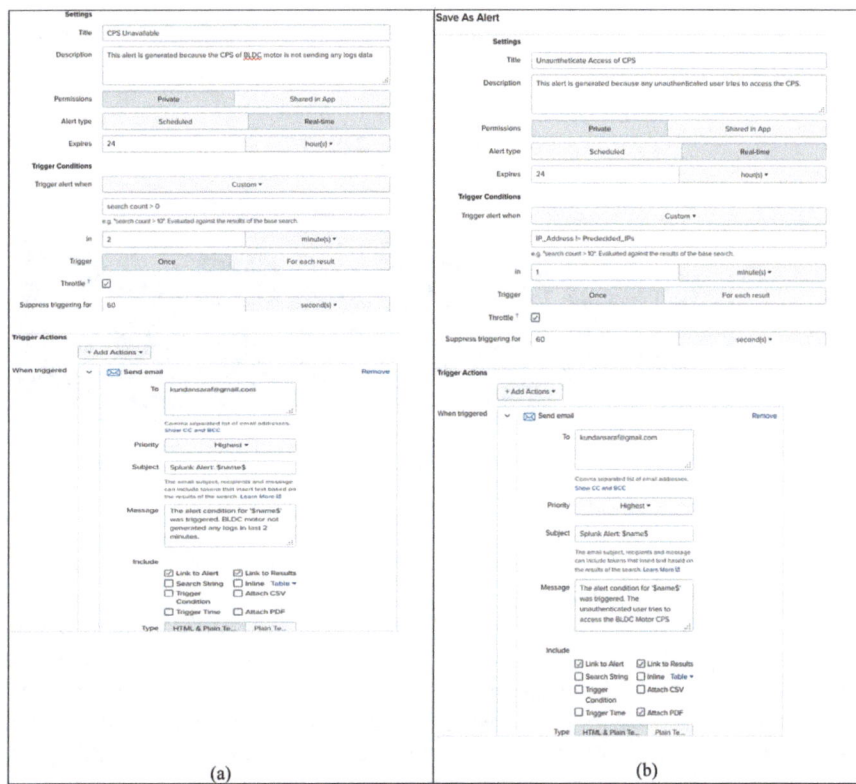

(a)   (b)

Fig. 8 Alerts created by Splunk, **a:**—Splunk Alert 1 created to inform the user for attack on availability of Cyber Physical System and **b** Splunk Alert 2 created to inform the user for attack on authenticity and authorization of Cyber Physical System

Table 5 BLDC motor speed and required time

Input Sequence followed by user: www.bldcmotor cps.in/remotecon trol/index.html	Reading from android application	Speed of BLDC Motor	Speed difference as compared to previous value	Sensing time
At Start	–	100 rpm	100 rpm	20 ms
First Input	250 rpm	249.90 rpm	250 rpm	130 ms
Second Input	750 rpm	750.01 rpm	500 rpm	160 ms
Third Input	200 rpm	199.96 rpm	550 rpm	161 ms
Fourth Input	1760 rpm	1760 rpm	1560 rpm	182 ms
Fifth Input	2040 rpm	2039.89 rpm	280 rpm	131 ms
Sixth Input	2668 rpm	2667.88 rpm	628 rpm	164 ms
Eighth Input	2993 rpm	2992.86 rpm	325 rpm	134 ms
Ninth Input	446 rpm	445.87 rpm	2547 rpm	194 ms

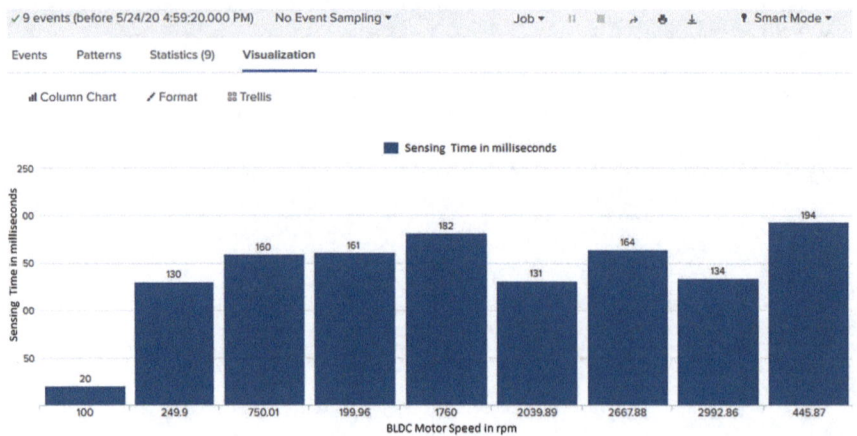

Fig. 9 Vertical bar chart of sensing time versus BLDC motor speed

given value to change the speed of BLDC motor. It also shows the time required to achieve a new user suggested speed by BLDC motor.

Figure 9 shows the vertical bar chart of sensing time versus BLDC motor speed. X-axis of this bar chart shows the speed in rpm. The Y-axis of this chart shows the time required to achieve the resulting speed in millisecond. Initially at the start of the motor, i t will takes a 100-rpm speed within a 20 ms. After the safe start of this motor, user can insert any value of speed less than the rated speed of motor, i.e., 3000 rpm proposed work. User has inserted nine different input values. It is observed that the time required to achieve the speed by motor after the new value inserted at webpage by user is less than 200 ms. This work uses the ultra-high-speed Internet. Hence, the time lag because of speed of Internet is very small, and it is included in 200 ms delay value of motor response. It can be observed by Table 5 and Fig. 9, the speed difference is directly proportional to time delay taken by BLDC motor to set its new user given value. It can be observed by second, third, and fourth input that, if the speed difference is 500 rpm to 630 rpm the time delay will be 160 ms to 164 ms. As observed by fourth and ninth input, if the user given speed difference is greater than 1500 rpm, the motor will take more time to settle down for new user given input. In this case, motor requires 182 ms and 194 ms to settle down for this speed values. In first, fifth and eighth input value if the speed difference is 250, 280, to 325 rpm, the motor will quickly settle down to its new user given speed value. In this case, motor takes 130 ms, 131 ms, and 134 ms time, respectively, to settle for its low difference speed value.

6   Conclusion

This system is mainly designed to remotely operate BLDC motor along with speed control feature. Hall Effect sensor and AVR microcontroller measure speed of BLDC motor without establishing actual contact with motor. Hence, this system reduces friction loss, increases accuracy, and improves efficiency as compared with conventional sensor which makes contact with motor shaft for speed measurement.

In this system, Wi-Fi controller ESP32-S2 is interfaced with AVR microcontroller. User can observe the present speed of BLDC motor and change its speed from remote location by accessing the webpage. User can also change the speed of motor by accessing this webpage. Hence, switching and speed changing operations can be performed by accessing the webpage.

Various security issues associated with this system are discussed in detail. The remedial measures for each issue are briefly explained. To secure this system against the attacks on confidentiality and integrity, it uses the Simon lightweight cryptographic algorithm. All log information generated by this system is transferred to Splunk Enterprise tool by configuring HTTP Event Collector. This tool performs the monitoring of log data received by CPS. Based on its monitoring, it creates an alert 1 indicating that the attack against availability of CPS. This system also configured to generate alert 2 indicating that the attack against authenticity and authorization. By limiting the number of users to use this webpage, this system is completely prevented against DoS and DDoS attack.

Future Scope
This work is specifically proposed for CPS of BLDC motor. It is essential to create system which can secure any CPS with multiple attacks [22]. The generalized system should be easily implemented in any CPS by inserting new few line code of HTTP Event Collector. One can also create a Splunk application which can be specifically used for securing the Cyber Physical System.

Resemblance of work with scope of book
This work is sensed the machine data, i.e., the logs of CPS and by reading these logs the Splunk create the appropriate alerts to protect the system by Cyber Attacks.

Splunk is mainly used in machine learning and optimized the real-time data. It creates the appropriate alerts based on the user requirements. These works perform the machine learning on CPS to detect the Cyber Attack

Funding
The total funding to build this Cyber Physical System has been invested by author of this publication.

Ethical Approval
This research maintains all the ethics of research.

Conflict of Interest
Both authors of this publication declare that they have no any conflict of interest for publication of this research.

Acknowledgements This research is held by using the 60 days trial version of Splunk Enterprise 8.0.3 software.

Declaration
This research does not perform experiment on any living thing such as human participant or animals.

Appendix

Code of the system

```
float adc,speed=0;

unsigned char txt[]="   ",temp[]="   ";

// LCD module connections

sbit LCD_RS at PORTD6_bit; // Allocate RS Pin of LCD to 6th Pin of Port D

sbit LCD_EN at PORTD7_Bit;

sbit LCD_D4 at PORTB0_bit;

sbit LCD_D5 at PORTB1_bit;

sbit LCD_D6 at PORTB2_bit;

sbit LCD_D7 at PORTB3_bit;

//Give direction to each pin of LCD

sbit LCD_RS_Direction at DDD6_bit;

sbit LCD_EN_Direction at DDD7_bit;

sbit LCD_D4_Direction at DDB0_bit;

sbit LCD_D5_Direction at DDB1_bit;

sbit LCD_D6_Direction at DDB2_bit;

sbit LCD_D7_Direction at DDB3_bit;

// End LCD module connections

char i;                    // Loop variable
```

```
void Move_Delay() {              // Function used for text moving

  Delay_ms(500);                 // You can change the moving speed here

}

void main(){

  //ADCON1 = 0b00000110;

  speed=0;

  ADC_Init();  // initialise ADC

  Lcd_Init(); // Initialize LCD

  Lcd_Cmd(_LCD_CLEAR); // Clear display

  Lcd_Cmd(_LCD_CURSOR_OFF); // Cursor off

          // Write text in first row

  Uart1_Init(9600);     // Baud Rate= 9600 means information is transferred in serial
port by 9600 bits per second

        Lcd_Out(1,1,"Contactless     "); // First Row starts with first column

        Lcd_Out(2,1," Tachometer     "); // Second Row starts with first column

        Delay_ms(2000); // Delay of 2 Second is given

        Lcd_Cmd(_LCD_CLEAR); // Clear the LCD initially

while(1)

      {

      Lcd_Out(1,1,"RPM:"); // Show RPM message on first row first column

      speed=0;

      adc=ADC_Read(0);  // Take no time read ADC means immediately

                      // read the ADC

      while(adc>512)   // When ADC value is greater than 512

        {

        adc=ADC_Read(0); // read the ADC

        Delay_ms(10);
```

```
        }
    adc=ADC_Read(0);

while(adc<200)       // When ADC value is less than 200 then

    {
    adc=ADC_Read(0);  // Read the ADC

    Delay_ms(1);

    speed=speed+1;

    }
    adc=ADC_Read(0);

while(adc>200)    // When ADC value is more than 200 then

    {
    adc=ADC_Read(0);  // Read the ADC

    Delay_ms(1);

    speed=speed+1;

    }
    speed=(60000/speed)/ 1.3; // Formula for actual speed of ADC

    IntToStr(speed,txt);    // Convert the integer into string

    Lcd_Out(2,-4,txt); // second column fourth row

    Delay_ms(1500);     // wait for 1.5 second

    while(1);

    }

}
```

References

1. H. Li, W. Li, H. Ren, "Fault-tolerant inverter for high-speed low-inductance BLDC drives in aerospace applications." IEEE Trans. Power Electron. **32**(3) (2017)
2. X. Song, J. Fang, B. Han, "High-precision rotor position detection for high-speed surface PMSM drive based on linear hall-effect sensors." IEEE Trans. Power Electron. **31**(7) (July 2016)
3. S. Sivaranjani, R. Rajeswari, Internet of things based industrial automation using brushless dc motor application with resilient directed neural network control FED virtual Z-source multilevel inverter topology. Wireless Pers. Commun. **102**(4), 3239–3254 (2018)

4. R. Saranya, R. Punithavalli, E. Nandakumar, R. Priya, "Web monitoring and speed control of solar based bldc motor with iot." In 2019 5th International Conference on Advanced Computing & Communication Systems (ICACCS), pp. 808–812. IEEE (2019)
5. M. Aqil, J. Hur, A direct redundancy approach to fault-tolerant control of BLDC motor with a damaged hall-effect sensor. IEEE Trans. Power Electron. **35**(2), 1732–1741 (2019)
6. M.S. Wang, S.C. Chen, C.H. Shih, Speed control of brushless DC motor by adaptive network-based fuzzy inference. Microsys. Technol. **24**(1), 33–39 (2018)
7. K.R. Saraf, M.P. Jesudason, "Encryption principles and techniques for the internet of things." In Cryptographic security solutions for the internet of things, pp. 42–66. IGI Global (2019)
8. Y. Liu, J. Zhao, M. Xia, H. Luo, Model reference adaptive control-based speed control of brushless DC motors with low-resolution Hall-effect sensors. IEEE Trans. Power Electron. **29**(3), 1514–1522 (2013)
9. Wu, Han-Chen, Min-Yi Wen, and Ching-Chang Wong. "Speed control of BLDC motors using hall effect sensors based on DSP." In 2016 International Conference on System Science and Engineering (ICSSE), pp. 1-4. IEEE, 2016
10. Scelba, Giacomo, Giulio De Donato, Mario Pulvirenti, Fabio Giulii Capponi, and Giuseppe Scarcella. "Hall-effect sensor fault detection, identification, and compensation in brushless DC drives." IEEE Transactions on Industry Applications 52, no. 2 (2015): 1542-1554
11. Kumpanya, Danupon, and Satean Tunyasrirut. "DSP-Based Speed Control of Brushless DC Motor." In Asian Simulation Conference, pp. 267-277. Springer, Berlin, Heidelberg, 2014
12. Wang, Wen-cheng. "A Motor Speed Measurement System Based on Hall Sensor." In International Conference on Intelligent Computing and Information Science, pp. 440-445. Springer, Berlin, Heidelberg, 2011
13. Q. Zhang, M. Feng, "Fast fault diagnosis method for hall sensors in brushless dc motor drives." IEEE Trans. Power Electron. (2018)
14. A. Herbadji, H. Goumidi, Y. Harbi, K. Medani, Z. Aliouat, "8 Blockchain for internet of vehicles security." Blockchain for Cybersecurity and Privacy: Architectures, Challenges, and Applications, **159** (2020)
15. Rasim Alguliyev, Yadigar Imamverdiyev, Lyudmila Sukhostat, Cyber-physical systems and their security issues. Comput. Ind. **100**, 212–223 (2018)
16. H.W. Lim, W.G. Temple, B.A.N. Tran, B. Chen, Z. Kalbarczyk, J. Zhou, Data integrity threats and countermeasures in railway spot transmission systems. ACM Trans. Cyber-Phys. Syst. **4**(1), 1–26 (2019)
17. Q. Gu, D. Formby, S. Ji, H. Cam, R. Beyah, Fingerprinting for cyber-physical system security: device physics matters too. IEEE Secur. Priv. **16**(5), 49–59 (2018)
18. A. Essa, T. Al-Shoura, A. Al Nabulsi, A.R. Al-Ali, F. Aloul, "Cyber physical sensors system security: threats, vulnerabilities, and solutions." In 2018 2nd International Conference on Smart Grid and Smart Cities (ICSGSC), IEEE (2018), pp. 62–67
19. Jacob Wurm, Yier Jin, Yang Liu, Hu Shiyan, Kenneth Heffner, Fahim Rahman, Mark Tehranipoor, Introduction to cyber-physical system security: a cross-layer perspective. IEEE Transactions on Multi-Scale Computing Systems **3**(3), 215–227 (2016)
20. S. Tanaka, K. Fujishima, N. Mimura, T. Ohashi, M. Tanaka, "IoT system security issues and solution approaches." Hitachi Review 65, no. 8, pp. 359–363 (2016)
21. Manual of Splunk Enterprise Getting Data In 8.0.3, 14th May 2020 https://docs.splunk.com/Documentation/Splunk/8.0.3/Admin/Howtousethismanual
22. P. Fu, J. Wang, X. Zhang, L. Zhang, R.X. Gao, Dynamic routing-based multimodal neural network for multi-sensory fault diagnosis of induction motor. J. Manuf. Syst. **55**, 264–272 (2020)

Optimization of Loss Function on Human Faces Using Generative Adversarial Networks

Vivek Bhalerao, Sandeep Kumar Panda, and Ajay Kumar Jena

Abstract Every human has a unique face that may vary with a wide variety of details from person to person. But, in twins, these differences are very less to be noticed but much significance. Hence, this raises a problem in creating digital faces of humans with all distinguishable features to that of a real human. Also, identify the faces that are not completely visible in a picture due to the distorted image or wrong angle of the face. To resolve this problem, in this research, we are building a model that can generate human after training of the models. For these face generation, we have used deep convoluted generative adversarial networks (DCGANs). This model is trained using the images available in the form OFA dataset that is being provided to the model. The images in the dataset are passed through a series of modifications called preprocessing. Next, our model is divided into three segments called generator, discriminator, and loss function optimization. These segments are trained individually and together form our complete model. The generator, generating the image of the human face and later, these images are provided to the discriminator. The discriminator that is also trained using the dataset starts discriminating against the image generated by the generator. The results are of generator, and discriminator is improvised using improvement analysis called loss function. Finally, and based on this result analysis, the generator is going to improve in its next epoch using the dataset image. As the loss values are noted for generator and discriminator, the values go as 1.1314, 0.8530, 0.8263, 0.8010 for the generator and 1.0013, 1.1288, 1.3122, 1.3111 for the discriminator. This shows a good loss reduction in generated images by a generator which means the enhancement in features of faces. The loss values increase in discriminator as it fails to discriminate the image. In the final stage of output, the generated face image is so real to be discriminated between fake and realistic face image. This model provides a solution to many existing problems of face recognition. This model can be used to identify people even if their face is partially visible on the screen by generating a clear image by a few modifications.

V. Bhalerao (✉) · S. K. Panda
Computer Science and Engineering Department, IcfaiTech (Faculty of Science and Technology), ICFAI Foundation for Higher Education (Deemed to be University), Hyderabad, Telangana, India

A. K. Jena
School of Computer Engineering, KIIT Deemed to be University, Bhubaneswar, Odisha, India

Keywords Deep convolutional generative adversarial network · Generator · Preprocessing · Optimization · Discriminator · Loss function

1 Introduction

The present world is growing toward a kind of lifestyle where daily life activities are easy and fast. People have been using and inventing various technologies to improve their lifestyles in a way they find it attractive and helpful. As referred in a research by Mitchel [1], the latest technology such as machine learning [2] is being implemented in almost all fields. Machine learning is a program to teach your computer. This learning is based on the model being coded basically in python language. The output of these models is based on the data provided to them. This data is provided to the model is called "dataset." These models are being improvised, and new ways of implementing the model are being brought into action. Although these technologies are being implemented, there have been issues with a computer being able to find accurate outputs of certain problems. One of these problems is the recognition of images.

There have been issues with image recognition that was captured from cameras, and these machine learning models gave very little accuracy. Objects present in the images have always been difficult to be identified. Different models like decision trees and others were implemented for obtaining better accuracy as cited in Mitchel in his research. However, these models were not able to provide the desired accuracy.

Dargan et al. [3] proposed an application that used a neural network for this problem. And this gave a higher accuracy when compared to other models. Neural networks [4] are a part of the deep learning model. The neural network was constructed in such a way that represents the working of neurons in the human brain. Also, it presents in application research by Affonse et al. [5] and Ashiquz-zaman et al. [6] data is processed through millions of neurons connected in our brain to make a decision. Similarly, data is processed through a lot of nodes and layers in neural networks to produce an output. Deep learning is a part of machine learning that provides output based on large amounts of data. This data is mostly unstructured, diverse, and interconnected. Deep learning models are mostly seen in chatbots applications. These chatbots are capable of answering almost all questions that a user can raise. And these varieties in question models, meaning, and other factors make it challenging for a program to satisfy the requirement, and thus, neural network which is trained using millions of questions was able to solve this problem.

Goodfellow et al. [7] propounded through an application that neural networks were improvised drastically such that, it started producing new objects such as images by combining the pieces of images that were based on the training done using the dataset. This generative model able to generate new images is called generative adversarial networks (GANs). Abhiroop et al. [8] proposed a project describing that GANs are of different types based on the way they are trained. Some of these are deep convoluted GANs (DCGANS), cycle GANS (CGANS) that are used in generating objects based on the dataset and type of output we desire. In this paper, we present you with a deep convoluted generative adversarial networks model that is going to produce the face images of people. The faces that are being generated here are also done using a huge dataset consisting of a lot of images.

The objective of this research is to generate human faces using GAN which has features the same as that of a real person. We have chosen deep convoluted GANs because it is a kind of object synthesizer that can be useful for this research over other types of neural networks. The deep convolution as in the name can be trained to give better results in the domain of generative outputs. The dataset that is being used is taken from an online source, and the dataset is called as "CelebA" dataset. The dataset is taken and preprocessed to make it ready to be trained by the model. This model is going to be consisting of building a generator, a discriminator, and next comes the part of the training of both the generator and discriminator separately. The output of generated images by the generator is going to be discriminated by the discriminator, and the accuracy of the image is measured using the loss function of both generator and discriminator separately. The loss is optimized using optimization techniques. And finally, we will be having a model that produces the desired output.

The model that is being used here is a TensorFlow-based. Hence, to be able to run the TensorFlow backend, we will be in need to use a good processor and a GPU with enough CUDA core for our model to run. The basic libraries that are being used are NumPy [9], TensorFlow [10], Matplotlib [11], OpenCV [12], and other few basic libraries as required.

The chapter was organized in seven sections. In Sect. 2, we describe the literature survey related to our work. In this section, we have gathered information about previous research done on the same domain. Next, in Sect. 3, we present the dataset that is being used in our research. In Sect. 4, we have proposed a novel model for generating faces for using the generative adversarial network. This includes steps, the methodology used, the input of data, and preprocessing which leads to the next Sect. 5, where we comprehending the building of three different models and training them accordingly. This model produces an output that is deliberated in the next section that is Sect. 6 of the chapter which is the result analysis. Finally, the conclusions and future directions were described in Sect. 7.

2 Literature Survey

Artificial intelligence is an area of computer science that deals with creating intelligent machines that can work and thinks like humans. To achieve the goals, researchers have found many techniques. One of these techniques that are being widely used in solving modern age problems is called machine learning. Machine learning mostly deals with training our machines in such a way that the results are predicted, and this prediction is made based on the data we provide it for training. The data that is going to be provided to us can be in various forms, organized or unorganized provided by a client or online source, and in few cases, we have to prepare a dataset consisting of the data that will be required to train our model.

In this machine learning, there are various algorithms, methods, and formulae to choose from. Each problem in machine learning has a unique approach to get the desired output. There is no particular solution available that can solve all the problems related to machine learning, and also, it is unpredictable that a particular model will surely give us the best output as required. Hence, for each problem, the trail method is applied with various algorithms and checks the accuracy in the output produced. And this accuracy value is used to decide the particular algorithm for a problem statement. In most of the cases, the result that is obtained is analyzed to understand the level o f model that is being used in a particular problem-solving. These analyses can make us understand the parameter value that is being used and maybe adjusting them may simply increase the accuracy rate of the model. As mentioned earlier, no parameter value or analysis is the same for different models. It varies with different models and needs to be found using different methods.

Machine learning algorithms are also used in problems dealing with image recognition. Although various machine learning algorithms were used, no solution provided the desired accuracy in the results. Later, the deep leaning model was used for image recognition problems. The deep learning model has become the most important, fascinating, and widely used method in computer vision. The deep learning model gave higher accuracy than other machine learning models. Deep learning uses neural networks as a model for producing the output. The architecture of neural networks is observed to be similar to that of neurons in the human brain. The neural network has nodes similar to that of neurons, and these are connected. The front propagation of these nodes is based on the bias or weights that are through the training of neural networks using the dataset. This finds the hidden patterns in the dataset and provides a better accuracy as the data provided to be mostly unstructured and widely varied. These neural networks are improvised in such a way that it can now generate new objects, and this is based on training done to it using the dataset called generative adversarial networks (GANs).

Hu et al. [13] and Isola et al. [14] work on computer vision presents GANs have various generative output applications such as text generation, images of face genera- tion, and sketch to realistic image generation. Faces are being generated using various methods such as face swap, manipulation of faces, attributes-based manipulation, and makeup-based face generation.

As proposed by Gatsy et al. [15], in one of their projects that deep convoluted generative adversarial network is a type of generative-based model that is being used in generating various types of objects such as numbers from a noised image of number and faces from pre-available faces based on the dataset provided. This model is built using "Keras" [16] interface powered by "TensorFlow" [17]. Keras is an API for high-level neural networks written in python language that can be used to train and build our models. TensorFlow is used for machine learning and is an open-source platform. The TensorFlow GPU is being used in the system for the execution of the neural network model.

Lesmana et al. [18] proposed in an application that the images that are present in the dataset are in the form of ".jpg". These cannot be directly inputted in the ".jpg" form in the python platform. Hence, converted into a numbered form that is processed by the model. These numbers are in the form of a matrix consisting of intensity values of three colors red, green, and blue (RGB) that range from 0 to 256 also preset in a project by Li et al. [19] This dataset is then modified as desired. This is similar to the way most of the machine learning models dataset is preprocessed. In image preprocessing, the modifications we usually do are cropping, normalization, resizing, grayscale, noise removal, or denoise, gray scaling.

Ulyanov et al. [20] and Madhuri et al. [21] proffered that the type of neural network that is used here is deep convoluted generative adversarial network. This means that the layers consisting of nodes in the network are not fully connected and uses convolutional strides for up-sampling and down-sampling. In a neural network, their hidden layers that are connected between the input and output layers that help in producing the accurate output. And in this model of a neural network, usage of "ReLU" [22] activation is made. The neural network that uses images as training takes images and splits it into smaller pieces of images with all the edges, curves, and corners. These factors of a complete image work as features of that particular image. For image identification, these features are used by the model. Also, based on these features, the new image is classified or identified by the discriminator. As discriminator also uses a neural network, the same methodology is used to adjust the weights to the nodes in the neural network. Similarly, the generator generates the image but the way of working of a generator is observed to be the opposite of that of working of a discriminator.

Kumar et al. [23] proposed the neural network used in the generator, once trained starts generating images, and the neural network in discriminator starts discriminating against the fake images that are being produced by a generator. The evaluation of these functions of generator and discriminator is done using loss function. The loss function is taken for both generator and discriminator separately. Finally, when all these factors are taken care of the loss being optimized for both of the sub-models, the output that is obtained consists of images of the faces with discrimination done based on the real images and the loss value generated for both generator and discriminator. To execute all the above sub-models to form a complete deep learning model, since the dataset used here is occupied 1 GB of memory, it can use a significant amount of time to load the data multiple times. Therefore, it is better to use the "pickle" [24] form of the loaded dataset. This function saves the data in the form of a python file that is efficient and easy to retrieve as presented by Zheng [25] in his conference.

The generation of faces is a quite fascinating problem. As the tiniest detail can change a whole person's face and be unrecognizable. But here the features of the face, depending on which the generator is trained, and it starts producing the faces, the face that is obtained may not consist of all recognizable features in the first epoch. Hence, as the epoch is carried on, the features of the image get more recognizable and similar to that of a real person.

3 Dataset

The dataset that is being used was collected at MMLAB at a Chinese University in Hong Kong. The dataset is named as "CelebA" dataset. This dataset consists of 202,599 images of various celebrities. This dataset is said to be having variations in posing, background, people, and rich annotations. With these images, we get 40 binary attributes for each image. Each image here is having a unique number to be differentiated easily. These images are present in the form of ".jpg". The dataset is available as present in the real world, and in the way, we normally need. Therefore, it requires modifications that can make the data be processed and trained the model easily. The images from the dataset are inputted to the python file using the "Open CV" library function. This library inputs the image in matrix form from a given path. This is later modified in this matrix form. This process of modifications is called preprocessing. Also, in this method of preprocessing, the features present in the data get highlighted by this process.

Here, in our research work, we used:

i.   TensorFlow 1.6.0 or above,
ii.   Numpy,
iii. Matplotlib, and
iv.   OpenCV.

A high-end configuration, GPU with multi-CUDA cores is to support TensorFlow API.

4 Proposed Method

The model that is going to be built is going to be consisting of different segments. Each segment here has its unique specifications, requirement, output and has almost equal impact on the complete model outcome. The basic methodology that is being used here is taking the dataset. Modifying the data as required. The building required sub-models. Then, we move toward the training and improving the accuracy part of the model. Since the accuracy is measured here is in terms of the loss function, optimization is done for the loss function for this model to improve accuracy. The basic method followed here is the modification of dataset such as resize, cropping, normalizing, and then feeding them to the machine learning model to train them and obtain results. These results are ameliorated using optimization techniques. Figure 1 shows the flow diagram of the proposed method that is being followed here.

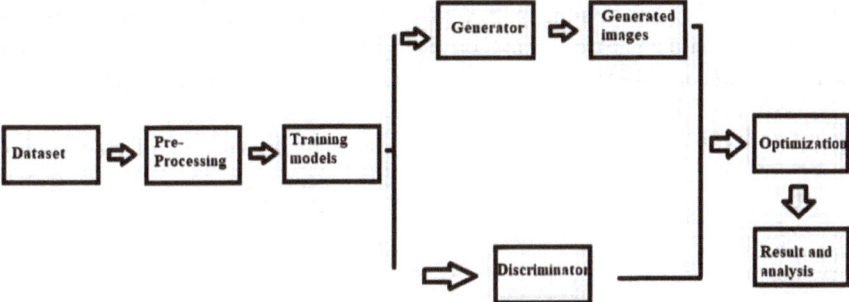

Fig. 1 Flow diagram of the proposed method

4.1 Inputting the Dataset

The dataset that is being used here consists of 202,599 images occupying almost 1 GB of data. To process this huge data, a lot of time is required. Hence, we try to reduce the size of the dataset by the methods of preprocessing discussed in Sect. 4.2. The images from the dataset are imported in the python platform in the form of a matrix consisting of pixel values. We have used the "OpenCV" library function to import our data from the dataset. OpenCV also called as "Open Source Computer Vision" library is a machine learning and computer vision library. An array is declared initially to store the values. OpenCV library function converts a single image into a matrix, and this is stored in the array. Hence, we are going to use a loop function to input all the images in matrix form and are appended into the initially declared array. Before the images are taken by the function, it requires an address to the location of the images. The path is declared using the "glob" library function as shown in Fig. 2.

Here "tqdm" library function was used to estimate the required time to load the dataset that can also be said to be the number of iterations per second. Three colors red, green, and blue (RGB) that are valued from 0 to 256 as per the color intensity at each pixel in the image is converted in the form of a matrix filled with these intensity

```
path=glob.glob("ADDRESS TO IMAGES/*.jpg")
image_from_dataset = []
for img in tqdm(path):
    n = cv.imread(img)
    image_from_dataset.append(n)

100%|██████| 202599/202599 [22:46<00:00, 148.28it/s]
```

Fig. 2 Inputting the dataset into python IDE

values. Figure 3 represents the matrix form of the image in Fig. 4 as it is imported and converted into an array in a python file. Figure 4 shows, when converted into a matrix have a shape of (218, 178, 3) that represents the height, width, and the number of colors that is "RGB".

```
[[[ 72   79 122]          [[ 81   81   81]
 [ 72   79 122]           [ 99   99   99]
 [ 72   78 123]           [108 108 108]
 ...                      ...
 [146 160 166]            [173 168 169]
 [217 217 217]            [164 159 160]
 [230 230 230]]           [164 159 160]]

[[ 64   73 117]           [[ 98   98   98]
 [ 65   74 118]           [109 109 109]
 [ 65   74 118]           [ 99   99   99]
 ...                      ...
 [149 163 169]            [150 145 146]
 [217 217 217]            [162 157 158]
 [231 231 231]]           [162 157 158]]

[[ 67   77 124]           [[ 98   98   98]
 [ 67   77 124]           [109 109 109]
 [ 68   78 125]           [100 100 100]
 ...                      ...
 [154 168 174]            [179 174 175]
 [218 218 218]    ~       [164 159 160]
 [232 232 232]]   ...     [164 159 160]]]]
```

Fig. 3 Matrix form of Fig. 4

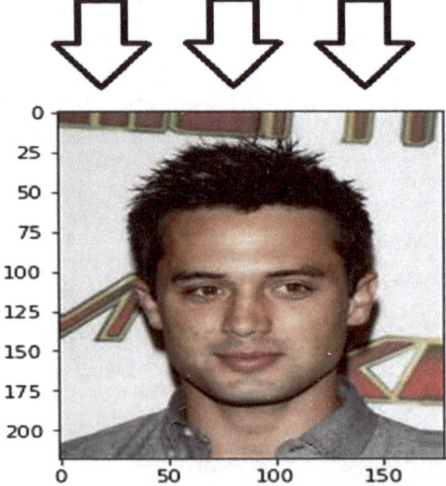

Fig. 4 Image in .jpg form

The images in the form of a matrix can be displayed as a real image using a matplotlib library using "show()" function. This function prints the image with length and width of the image as the *y*-axis and *x*-axis, respectively. Here all three colors intensity is mixed together that makes a perfect colored image. After all the images are imported into python IDE, these are later moved toward the steps of preprocessing.

4.2 Preprocessing

The preprocessing is the basic and most important step in building a model. It is a way of preparing our data that is going to be used for training our model. Image preprocessing can be done using various methods like resizing, reshaping, grayscale, segmentation, normalization, and many more. Here, we have used a few methods for preprocessing of the images in the dataset. The images available here need to be called individually. Hence, the provision was made to all the images with a unique batch number to be called separately and processing is done on it. Here, three preprocessing steps were used, and later these are spilled into training and testing data. The training data will be used to train the neural network of a generator model to generate fake images, and the testing data is used to train the discriminator that is going to discriminate fake images.

4.2.1 Crop

The images we have consist of extra space of unwanted image area. This area will reduce the efficiency of the model as it occupies extra memory to that of a required image, and also, the face is exactly not being highlighted. This unwanted area is removed by cropping the image, and we get the face focused area of the image. Figure 6 shows the cropped image of Fig. 4, and the blue color indicates that only the blue pixel intensity of the image is being read and showed by the python library. The colors RGB need to be modified to get the final cropped image. And the cropping can be done by removing the particular length from all sides of the image. For example, the image in Fig. 4 has length and width is 218 and 178, respectively. But the face has occupied the length from 25 to 175 and width from 25 to 125. This means that we can remove a length of 25 from every side. So after cropping the image the final image, we get is Fig. 6. And the cropping can be done using a simple line of code.

Fig. 5 Cropping the image

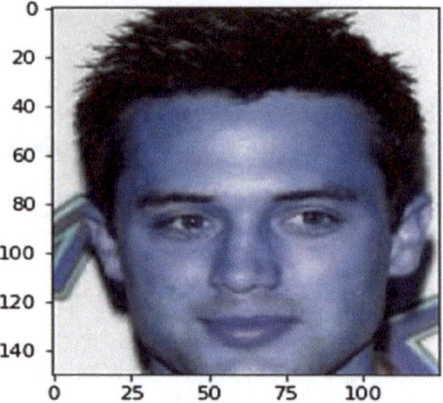

Fig. 6 Cropped images with only blue color intensities

In Fig. 5, we are removing the height by 25 units and width by 25 units on both sides. This will remove the particular length of the image from a single image. Therefore, a loop to call all images and cropping is done to every image (Fig. 6).

4.2.2 Normalization

Normalization is a method in which an image is a process of converting the pixel density range to the desired range. The pixel intensity values range from 0 to 255. This is altered to reduce this range using the following formula given in Eq. 1.

$$I_N = (\text{newMax} - \text{newMin})\frac{1}{1 + e^{-\frac{I-\beta}{\alpha}}} + \text{newMin} \tag{1}$$

This formula is used for the nonlinear type of normalization and follows a sigmoid function. Here, the "newMax" and "newMin" are the range of intensities up to which they need to be altered and α, β are the input intensity range and up to which the new intensity range should be centered. OpenCV function provides us the facility to normalize the image. Therefore, here we can use this function to normalize the image by setting α and β values. Figure 8 shows the normalized image of Fig. 6, and it can be done using the code as shown in Fig. 7.

```
norm_image = cv2.normalize(ima, None, alpha=-0.5, beta=0.5, norm_type=cv2.NORM_MINMAX, dtype=cv2.CV_32F)
```

Fig. 7 Normalization of image

We have set α and β values to 0.5, and this is done for a single image as shown in Fig. 7. To normalize all the images, just apply for loop and call images one by one and normalize them. Normalization type is "cv2.NORM_MINMAX" and datatype object declared here is "cv2.CV_32F" (Fig. 8).

As stated, that density values are altered, this means that, as the original image densities range from 0 to 256, all these density values are changed to a particular range that is smaller than the original range. And this method will either increase or decrease the intensities of the color at each pixel as required.

4.2.3 Resize

Resizing image makes the image easy for training the model as it reduces the size of the image. As, we can observe, in Fig. 9 that after the image is resized, the image starts looking a bit blurred. This is because the pixel size is reduced, this means that the original size of the image being (218, 178) is reduced to (50, 50).

Fig. 8 Normalized image of the dataset

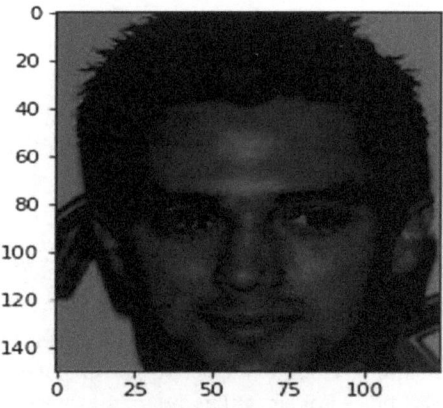

Fig. 9 Image after resizing

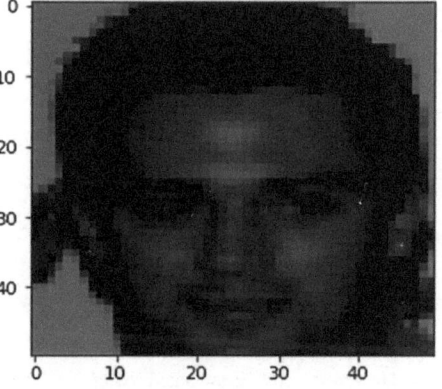

```
resized = cv.resize(norm_image, (50,50), interpolation = cv.INTER_AREA)
```

Fig. 10 Resizing the image

Resizing is also done using the OpenCV library function using the following line of code shown in Fig. 10. The dimension is set to (50, 50) that is shown in Fig. 10 that is the length and height of the image. The interpolation selected here is "Opencv.INTER_AREA".

The images that are present in the dataset are preprocessed and now all images are in a similar format to that of the image present in Fig. 9. With these modifications, we end the preprocessing method on the images present in the dataset and move further toward the building and training our model. The dataset is next split into two parts that is one for training our model and the second part for testing our model.

5   Building the Model

Before starting to build our model, we need to either build a complete neural network or use a pre-build neural network available as API. "Keras" provides us with the required neural network API. Keras is an API for high-level neural network written in python language, and it uses "TensorFlow" in the backend that is an open-source platform for machine learning. Here, we have used Keras from TensorFlow GPU for neural networks. Install the TensorFlow Function library in your python IDE. Now, build a function to call TensorFlow.

Here, TensorFlow is imported with a variable "tf". The "placeholder" is for assigning a memory space for the data we are going to store as observable in Fig. 11. This place holder will assign memory for every image feature in the neural network

```
import tensorflow as tf



real_input = tf.placeholder(tf.float32, shape=(None, width, height, channels), name='input_real')
z_input = tf.placeholder(tf.float32, (None, z_dim), name='input_z')
rate = tf.placeholder(tf.float32, name='learning_rate')
```

Fig. 11 Using TensorFlow background

at the nodes. By this, all the nodes are consisting of all the required features that can be assigned during the training of the model. And the arguments in real_input are the values of width, height, and channels that are 3. The arguments in z_input are the dimensions of the image. The learning rate is for a neural network. This is declared as a function bypassing, the parameter such as height, width, and channels of the image.

5.1 Discriminator

Next, we start building the sub-model called as the discriminator. A discriminator is a neural network used to discriminate between real and fake images. It is trained using real images from the dataset. It is built using three layers whose number of nodes per layer increases as the move from input layers to output layers of the neural network. First, we set the first layer and make sure the "leaky ReLu" function value is obtained.

Here, the 2D convolutional layer was built with 64 filters, kernel size as 5, and strides as 2 (Fig. 12). And the second line gives us the maximum of the arguments present inside it which is the leaky ReLu. The next layer as shown in Fig. 13 is built with 256 filters, and batch normalization is done to make the network work faster.

In layer 2, batch normalization was done including leaky ReLu. Further, we move forward to the third layer that consists of 512 filters. Figure 14 shows the declaration of layer three with normalization of the images.

Next, it is flattened that converts the feature map of the connected network into a single column. Then logits are passed that are the inputs to the layers and output is obtained. We have used the deep convoluted neural network (DCGAN) for this

```
layer1 = tf.layers.conv2d(images, 64, 5, 2, 'SAME')
relu1 = tf.maximum(alpha * conv1, conv1)
```

Fig. 12　Declaring the first layer

```
layer2 = tf.layers.conv2d(lrelu1, 128, 5, 2, 'SAME')
normalization = tf.layers.batch_normalization(layer2, training=True)
relu2 = tf.maximum(alpha * normalozation2, normalization2)
```

Fig. 13　Second layer of the network

```
layer3 = tf.layers.conv2d(lrelu2, 256, 5, 1, 'SAME')
normalisation3 = tf.layers.batch_normalization(layer3, training=True)
relu3 = tf.maximum(alpha * normalisation3, normalisation3)
```

Fig. 14 Third layer of the network

model. Further, rectified linear unit (ReLU) activation is being used for this model. Figure 15 displays the brief idea of the working of a discriminator.

A discriminator works opposite to that of a generator. A discriminator is inputted with real images that are the images from the dataset and trained with it such that, as the image from the generator is provided as an input to it, the discriminator discriminates it with either fake or real image. Here, to discriminate an image, it splits the image and looks at its features such as edges, corners, and other such objects. Then, this is taken from all such images, and the same is done with the generated image and finally discriminated between real or fake images.

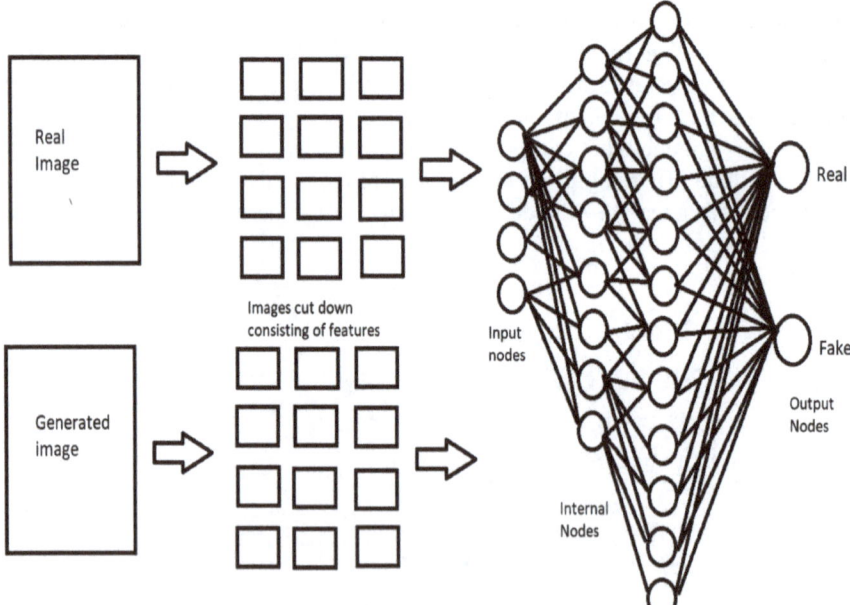

Fig. 15 Brief descriptions about the working of the discriminator

5.2 *Generator*

A generator works opposite that of the discriminator. The generator is trained with the images from the dataset. Next, it generates the images as output. It works by generating small images of edges and corners that are the features of the image and then joining them to form a complete image. This generator is also of three layers. As the path is moved in a forward direction the number of filters decreases. The leaky ReLu values obtained from every layer will the first layer consists of 512 filters as shown in Fig. 16 with batch normalization of images.

The second layer of the neural network is similar to 256 filters. As observable in Fig. 17, the second layer is formed with required filters, normalization is done for the nods and leaky ReLu value is also taken from this layer.

And the final layer in the network consists of 128 filters with normalization for the faster working of neural networks as shown in Fig. 18. The values for all the ReLu functions for every layer is taken separately and later used.

Later, logits are added to the generator and the output is taken. This image is then provided to a discriminator for discrimination. Figure 19 shows a brief idea of the working of a generator.

A discriminator tries to discriminate between a real and fake image provided by the generator and a generator tries to fool the discriminator by producing perfect images by improving itself.

Fig. 16 First layer declaration in our generator

```
layer1 = tf.layers.dense(z, 2*2*512)
normalisation = tf.layers.batch_normalization(layer1, training=is_train)
relu2 = tf.maximum(alpha * normalisation, normalisation)
```

Fig. 17 Second layer of the network

```
layer2 = tf.layers.conv2d_transpose(lrelu2, 256, 5, 2, padding='VALID')
normalisation = tf.layers.batch_normalization(layer2, training=is_train)
relu = tf.maximum(alpha * normalisation, normalisation)
```

Fig. 18 Third layer of the network

```
layer3 = tf.layers.conv2d_transpose(relu, 128, 5, 2, padding='SAME')
normalisation = tf.layers.batch_normalization(layer3, training=is_train)
relu2 = tf.maximum(alpha * normalisation, normalisation)
```

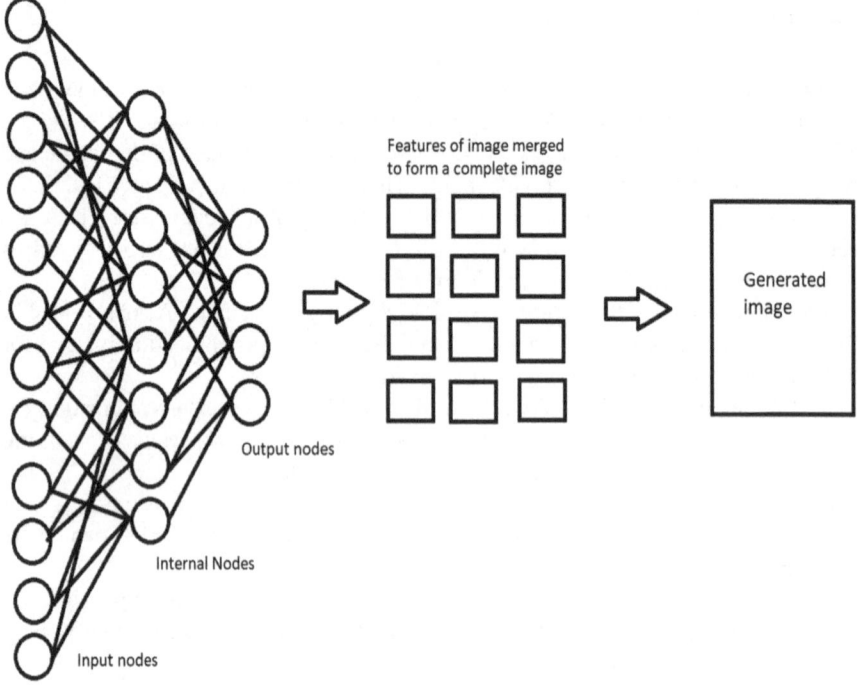

Fig. 19 Brief description of the working of the generator

5.3 Training of the Sub-models

Now, the training of these build sub-models is done. The image data that is prepro-cessed is provided to both discriminator and generator for training. Then based on this training, the generator starts generating images and the discriminator differentiates between the fake images.

5.4 The Loss Function

The loss function is the deviation of the accuracy line from the actual value-line when plotted on a graph. A loss value of zero means that the model used is a perfect model that is next to impossible to build. The loss function has no unit but rather it is just the difference of actual value to obtained value. The loss function is found for all the sub-models of the complete model. That is for the generator, discriminator separately. The loss function for a generator is based upon the real images provided to it. And for that of a discriminator, it is based upon the fake images. The loss of the generator and discriminator reduces with progression in epochs. For finding the loss, we use the sigmoid loss function. Sigmoid loss function also called binary cross-entropy loss is used here because the loss calculated by it for our convoluted neural network for a class is independent of another class. This means we do not need to look at multiple classes for while analysis. Hence, it is used with a desire to obtain better accuracy. This function can also be used for multi-label classification for better accuracy. This loss function is given by Eq. 2.

$$(1 - z) * x + 1 * (\log(1 + \exp(-\text{abs}(x))) + \max(-x, 0)) \tag{2}$$

Here x is the values of input to the generator and discriminator and z is the label of the image. This is later optimized by arranging the values of the parameters. For finding the loss, inputs from the models are taken and find the loss for the real images first as observable in Fig. 20.

Then, the loss of fake images or images produced by the generator is found. By giving the logits that were obtained during the production of fake images. Here, the sigmoid cross-entropy function was used as shown in Fig. 21. The values of these losses are calculated using the above formula.

Later the values from the loss function are optimized. For optimization of this loss function, first, the variables for generator and discriminator are declared. Next, the values are optimized using the pre-build function. In Fig. 22, we are optimizing the values of the loss function of generator and discriminator networks that were obtained before and an exceptional handler "with" was also used to make the code more readable and efficient.

In the above code, the "Adam optimizer" function was used and the loss function values are also minimized for both discriminator and generator.

```
real_loss = tf.reduce_mean(
    tf.nn.sigmoid_cross_entropy_with_logits(logits=real_logits,
                                labels=tf.ones_like(real_images) * label_smoothing)
```

Fig. 20 Loss of real images

```
fake_loss = tf.reduce_mean(
    tf.nn.sigmoid_cross_entropy_with_logits(logits=fake_logits,
                                labels=tf.zeros_like(fake_image_model)))
```

Fig. 21 Loss of fake image

```
with tf.control_dependencies(tf.get_collection(tf.GraphKeys.UPDATE_OPS)):
    discriminator_optimisation = tf.train.AdamOptimizer(dis_rate, beta1=beta1).minimize(dicriminagor_loss, var_list=d_vars)
    generator_optimisation = tf.train.AdamOptimizer(lear_rate, beta1=beta1).minimize(generator_loss, var_list=g_vars)
```

Fig. 22 Optimization of the loss function

6 Result Analysis

As multiple epochs are carried on the generator emerges toward producing a perfect image. The generator that tries to fool the discriminator with its fake images evolves toward producing images that are discriminated against as a real image by the discriminator. At this point, where the generator and discriminator move toward the perfect model, it is called an equilibrium state. At this point, the loss of the generator is very low. As the loss of generator is based on the real images, it converges toward almost zero and the discriminator whose loss is based on fake images also moves toward almost zero. The following series of figures that are 23(a), (b), (c), (d) shows the generated images by the generator, and the loss obtained using the loss function is shown in Table 1. To view the output, the images from the generator are displayed on the screen using the "matplotlib.pyplot.imshow()" function. With the loss function, values printed above it.

Figure 23a is the first generated image with a loss of 1.1314 and 1.0013 as generator and discriminator loss, respectively, and does not consist of any identical feature to that of a face. Moving toward the next images that are Fig. 23b which is also not identical, the generator produces an image (Fig. 23c) that is a bit identical and the final image (Fig. 23d) has the loss values 0.8010 and 1.3111 as generator and discriminator loss values, respectively, has almost all identical features to that of a face such as ears, eyes, nose, and mouth. Hence, the loss values are decreasing as the images are getting generated and the perfection in the images is increasing. This means the discriminator loss is increased as it gets failed in discriminating image as a fake one.

Table 1 Loss of values of images in Fig. 23 from (a) to (d) in a respective manner

Figure number	23(a)	23(b)	23(c)	23(d)
Generator loss	1.1314	0.8530	0.8263	0.8010
Discriminator loss	1.0013	1.1288	1.3122	1.3111

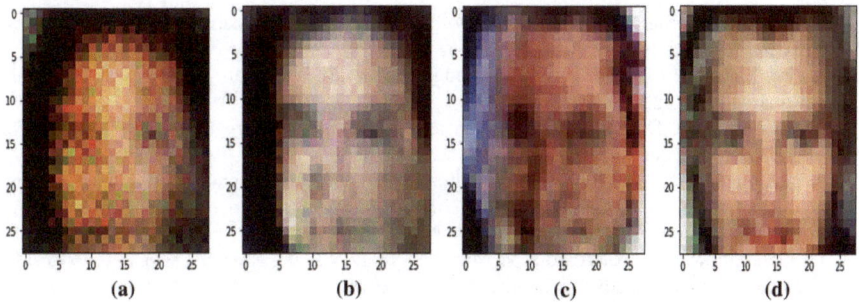

Fig. 23 **a** Image generated in the first epoch by the generator with generator loss as 1.1314 and discriminator loss as 1.0013. **b** Image generated in first epoch by the generator with generator loss as 0.8530 and discriminator loss as 1.1288. **c** Image generated in second epoch by the generator with generator loss as 0.8263 and discriminator loss as 1.3122. **d** Image generated in second epoch by the generator with generator loss as 0.8010 and discriminator loss as 1.3111

7   Conclusion and Future Work

In our research work, we were able to build a deep convoluted generative adversarial network that generates a perfect image of faces of people. That is not discriminated against as fake image by discriminator and these generated images look similar to that of a real image. As the epochs are carried out the loss function of the images generated by the generator is reduced. Further, the application of different types of GANs can be made and also obtain more accuracy. And its applications can vary from an attendance system to criminal's face recognition. Whereas, GANs can be used for the creation of animation, sketch to real image conversion, and also have many other applications. As the GANs used here are only DCGANs, other types of GANs can be used as per the desired output by the user. Neural networks require a lot of data to train to obtain better accuracy, the dataset provided to us is also huge and perfect for training our model. Since the dataset is huge, it is better to use the pickle function to store the data as a python file, such that there are no issues with the loading of the dataset.

In this chapter, we have tried to explore some of the unexplored areas related to GANs for face image generation. This helps us understand neural networks using an application basis. Further work on this can help us achieve results in various contexts such as image recognition and classification.

References

1. T.M. Mitchel, *The Discipline of Machine Learning*. CMU-ML-06-108 (2006)
2. V. Blanz, T. Vetter, A morphable model for the synthesis of 3D faces, in *Proceedings of 26th Annual Conference on Computer Graphics and Interactive Techniques*, Los Angeles (1999)

3. S. Dargan, M. Kumar, M.R. Ayyagari, G. Kumar, A survey of deep learning and its applications: a new paradigm to machine learning (2019)
4. J. Zhu, T. Park, P. Isola, A. Efros, Unpaired image-to-image translation using cycle-consistent adversarial networks. ArXiv (2017)
5. C. Affonso, A.L.D. Rossi, F.H.A. Vieria, A.C.P.D.L.F.D. Carvalho, Deep learning for biological-image classification. Expert Syst. Appl. **85** (2017)
6. A. Ashiquzzaman, A.K. Tushar, Handwritten Arabic numeral recognition using deep learning neural networks, in *Proceedings of IEEE International Conference on Imaging, Vision & Pattern Recognition* (2017)
7. I.J. Goodfellow, J. Pouget-Abadie, M. Mirza, B. Xu, D. Warde-Farley, S. Ozair, A.C. Courville, Y. Bengio, Generative adversarial nets, in *NIPS* (2014)
8. K.A. Tejomay, K.K. Kamarajugadda, Sketch to photo conversion using cycle-consistent adversarial networks (2020)
9. A. Anuar, K.M. Saipullah, N.A. Ismail, Y. Soo, OpenCV based real-time video processing using an android smartphone. IJCTEE **1**(3)
10. C. Gupta, N.S. Gill, Machine learning techniques and extreme learning machine for early breast cancer prediction (2020)
11. A. Choromanska, M. Henaff, M. Mathieu, G.B. Arous, Y. LeCun, The loss surfaces of multilayer networks, in *AISTATS* (2015)
12. W.M. Czarnecki, R. Jozefowicz, J. Tabor, Maximum entropy linear manifold for learning discriminative low-dimensional representation, in *Joint European Conference on Machine Learning and Knowledge Discovery in Databases* (Springer, 2015)
13. J. Hu, L. Shen, G. Sun, Squeeze-and-excitation networks, in *2018 IEEE/CVF Conference on Computer Vision and Pattern Recognition* (2017), pp. 7132–7141
14. P. Isola, J. Zhu, T. Zhou, A.A. Efros, Image-to-image translation with conditional adversarial networks, in *IEEE Conference on Computer Vision and Pattern Recognition (CVPR)* (2017), pp. 5967–5976
15. L.A. Gatys, A.S. Ecker, M. Bethge, Image style transfer using convolutional neural networks, in *2016 IEEE Conference on Computer Vision and Pattern Recognition (CVPR)* (2016), pp. 2414–2423
16. L. Tran, X. Yin, X.M. Liu, Disentangled representation learning GAN for pose-invariant face recognition, in *Proceedings of the IEEE Conference on Computer Vision and Pattern Recognition (CVPR)*, Honolulu (2017)
17. J.C. Principe, D. Xu, J. Fisher, Information-theoretic learning. Unsupervised adaptive filtering (2000)
18. S.B. Lesmana, E. Suhartanto, A. Suharyanto, V. Dermawan, Geospatial and artificial neural network applications for prioritization of watershed prediction (2020)
19. Y.J. Li, S.F. Liu, J.M. Yang, M.H. Yang, Generative face completion, in *Proceedings of IEEE Conference on Computer Vision and Pattern Recognition (CVPR)*, Honolulu (2017)
20. D. Ulyanov, A. Vedaldi, V.S. Lempitsky, Instance normalization: the missing ingredient for fast stylization. ArXiv, abs/1607.08022 (2016)
21. D. Krishna Madhuri, R.V.V.S.V. Prasad, A ML and NLP based framework for sentiment analysis on bigdata (2020)
22. S. Reed, Z. Akata, X.C. Yan, L. Logeswaran, B. Schiele, H. Lee, Generative adversarial text to image synthesis, in *Proceedings of 33rd International Conference on Machine Learning*, New York (2017)
23. K.S. Kumar, G.R. Chandra, D. Sukheja, Cotton disease detection using deep learning (2020)
24. T.M. Mitchell, *Machine Learning* (McGraw-Hill International, 1997)
25. Z.D. Zheng, L. Zheng, Y. Yang, Unlabeled samples generated by GAN improve the person re-identification baseline in vitro, in *Proceedings of IEEE International Conference on Computer Vision (ICCV)*, Honolulu (2017)